DEPARTMENT OF THE ARMY
U.S. Army Corps of Engineers
Washington, DC 20314-1000

EM 1110-2-1420

CECW-EH-Y

Manual
No. 1110-2-1420

31 October 1997

Engineering and Design
HYDROLOGIC ENGINEERING REQUIREMENTS
FOR RESERVOIRS

1. Purpose. This manual provides guidance to field office personnel for hydrologic engineering investigations for planning and design of reservoir projects. The manual presents typical study methods; however, the details of procedures are only presented if there are no convenient references describing the methods. Also, publications that contain the theoretical basis for the methods are referenced. Many of the computational procedures have been automated, and appropriate references are provided.

2. Applicability. This manual applies to all HQUSACE elements and USACE commands having civil works responsibilities.

FOR THE COMMANDER:

OTIS WILLIAMS
Colonel, Corps of Engineers
Chief of Staff

CECW-EH-Y	**DEPARTMENT OF THE ARMY** **U.S. Army Corps of Engineers** **Washington, DC 20314-1000**	EM 1110-2-1420

Manual
No. 1110-2-1420

31 October 1997

Engineering and Design
HYDROLOGIC ENGINEERING REQUIREMENTS
FOR RESERVOIRS

Table of Contents

Chapter 1
Introduction

1-1. Purpose

This manual provides guidance to field office personnel for hydrologic engineering investigations for planning and design of reservoir projects. The manual presents typical study methods; however, the details of procedures are only presented if there are no convenient references describing the methods. Also, publications that contain the theoretical basis for the methods are referenced. Many of the computational procedures have been automated, and appropriate references are provided.

1-2. Applicability

This manual applies to all HQUSACE elements and USACE commands having civil works responsibilities.

   a. Scope. This manual provides information on hydrologic engineering studies for reservoir projects. These studies can utilize many of the hydrologic engineering methods described in the manuals listed in paragraph 1-4. Hydraulic design of project features are not included here; they are presented in a series of hydraulic design manuals.

   b. Organization. This manual is divided into four parts. Part 1 provides basic hydrologic concepts for reservoirs. Reservoir purposes and basic hydrologic concerns and methods are presented. Part 2 describes hydrologic data and analytical methods. Part 3 covers storage requirements for various project purposes, and the last, Part 4, covers hydrologic engineering studies.

1-3. References

Required and related publications are listed in Appendix A.

1-4. Related H&H Guidance

   a. Engineer manuals. This engineer manual (EM) relies on, and references, technical information presented in other guidance documents. Some of the key EM's for reservoir studies are listed below. Additionally, there are related documents on hydraulic design for project features associated with reservoir projects. This document does not present hydraulic design concepts.

   EM 1110-2-1201   Reservoir Water Quality Analysis

   EM 1110-2-1415   Hydrologic Frequency Analysis

   EM 1110-2-1416   River Hydraulics

   EM 1110-2-1417   Flood-Runoff Analysis

   EM 1110-2-1602   Hydraulic Design of Reservoir Outlet Works

   EM 1110-2-1603   Hydraulic Design of Spillways

   EM 1110-2-1701   Hydropower

   EM 1110-2-3600   Management of Water Control Systems

   EM 1110-2-4000   Sedimentation Investigation of Rivers and Reservoirs

These manuals provide the technical background for study procedures that are frequently required for reservoir analysis. Specific references to these EM's are made throughout this document.

   b. Engineer regulations. There are several engineer regulations (ER) which prescribe necessary studies associated with reservoir projects. The most relevant ER's are listed below.

   ER 1110-2-1460   Hydrologic Engineering Management

   ER 1110-2-7004   Hydrologic Analysis for Watershed Runoff

   ER 1110-2-7005   Hydrologic Engineering Requirements for Flood Damage Reduction Studies

   ER 1110-2-7008   Hydrologic Engineering in Dam Safety

   ER 1110-2-7009   Hydrologic Data Collection and Management

   ER 1110-2-7010   Local Protection - Safety/ Workability

   ER 1110-8-2(FR) Inflow Design Floods for Dams and Reservoirs

These and other regulations should be consulted prior to performing any hydrologic engineering study for reservoirs. A current index of regulations should be consulted for new and updated regulations.

PART 1

HYDROLOGIC ENGINEERING CONCEPTS
FOR RESERVOIRS

Chapter 2
Reservoir Purposes

2-1. Congressional Authorizations

a. *Authorization of purposes.* The United States Congress authorizes the purposes served by U.S. Army Corps of Engineers reservoirs at the time the authorizing legislation is passed. Congress commonly authorizes a project "substantially in accordance with the recommendations of the Chief of Engineers," as detailed in a separate congressional document. Later, additional purposes are sometimes added, deleted, or original purposes modified, by subsequent congressional action. When the original purposes are not seriously affected, or structural or operational changes are not major, modifications may be made by the Chief of Engineers (Water Supply Act 1958).

b. *General legislation.* Congress also passes general legislation that applies to many projects. The 1944 Flood Control Act, for example, authorizes recreational facilities at water resource development projects. This authority has made recreation a significant purpose at many Corps reservoirs. Similar general legislation has been passed to enhance and promote fish and wildlife (1958) and wetlands (1976). The Water Resource Development Act of 1976 authorizes the Chief of Engineers, under certain conditions, to plan and establish wetland areas as part of an authorized water resource development project. A chronology of the congressional legislation authorizing various purposes and programs is shown in Figure 2-1 (USACE 1989).

c. *Additional authorization.* Figure 2-1 illustrates how additional authorizations have increased the number of purposes for which the Corps is responsible both in planning and managing water resource development projects. The first authorizations were principally for navigation, hydroelectric power, and flood control. Later authorizations covered a variety of conservation purposes and programs. During drought when there is a water shortage, all purposes compete for available water and are affected by the shortage. The more purposes and programs

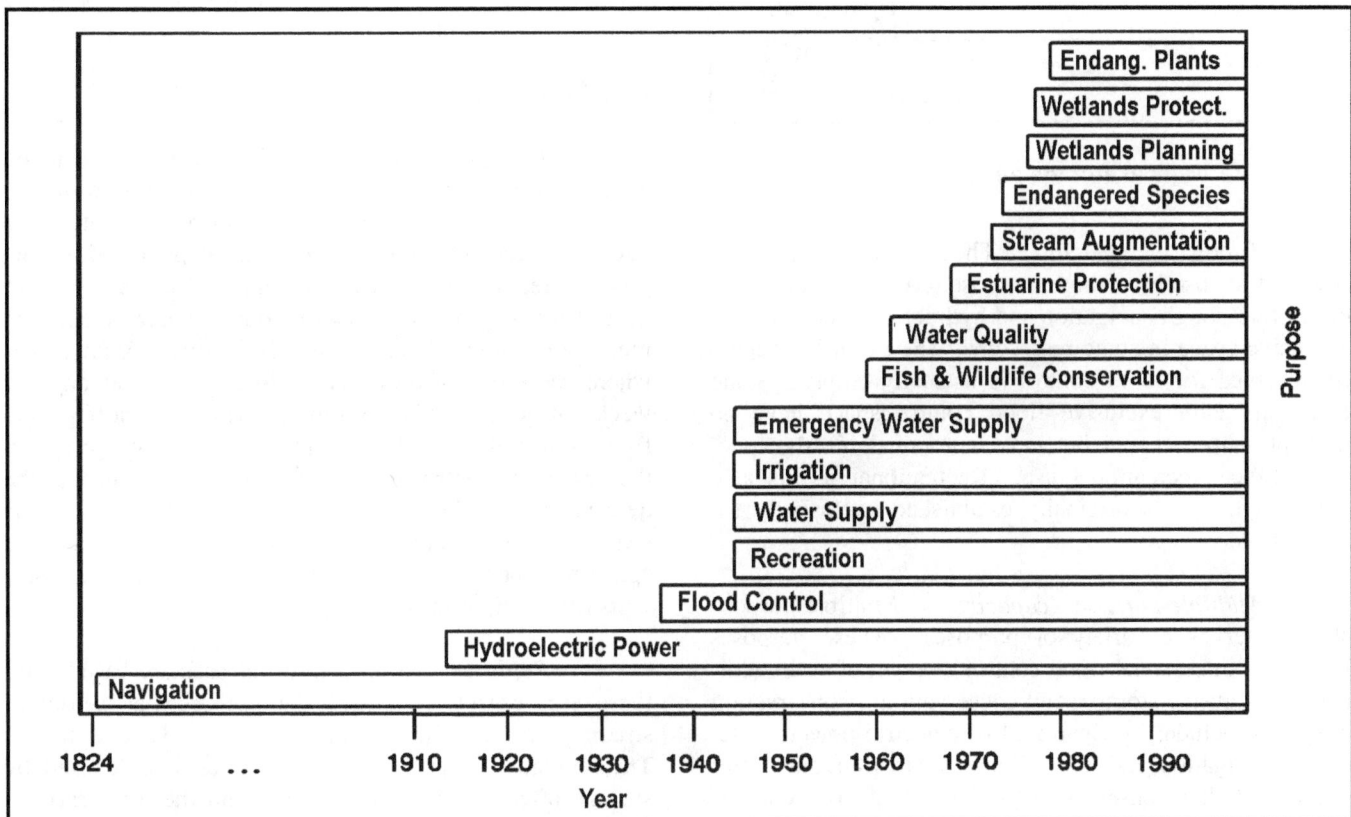

Figure 2-1. Purposes and programs authorized by Congress

there are to serve, the greater the potential for conflict, and the more complex the task of managing existing supplies. "Authorized and Operating Purposes of Corps of Engineers' Reservoirs" (USACE 1992) lists the purposes for which Corps operated reservoirs were authorized and are operated.

2-2. Reservoir Purposes

a. Storage capacity. A cross section of a typical reservoir is shown in Figure 2-2. The storage capacity is divided into three zones: exclusive, multiple-purpose, and inactive. While each Corps reservoir is unique both in its allocation of storage space and in its operation, the division of storage illustrated by Figure 2-2 is common.

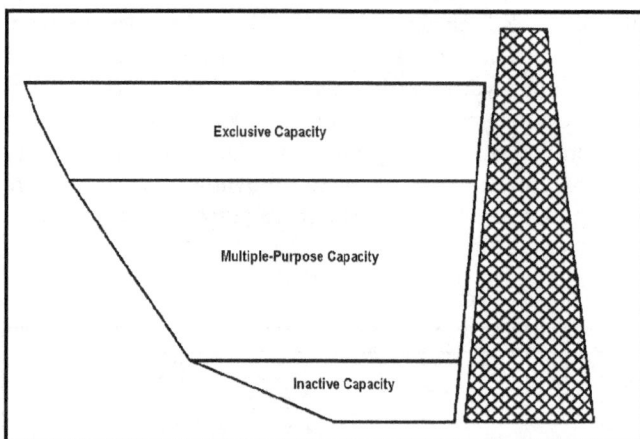

Figure 2-2. Typical storage allocation in reservoirs

b. Exclusive capacity. The exclusive space is reserved for use by a single purpose. Usually this is flood control, although navigation and hydroelectric power have exclusive space in some reservoirs. The exclusive capacity reserved for flood control is normally empty. Some reservoirs with exclusive flood control space have no multiple-purpose pool but have a nominal inactive pool that attracts recreational use. Recreational use is also common on pools originally established exclusively for navigation.

c. Multiple-purpose capacity. Multiple-purpose storage serves a variety of purposes. These purposes include both seasonal flood control storage, often in addition to exclusive storage, and conservation. Conservation purposes include: navigation, hydroelectric power, water supply, irrigation, fish and wildlife, recreation, and water quality. Other conservation purposes such as wetlands, groundwater supply and endangered species, while not

included in this manual, are nonetheless important in water control management.

d. Inactive capacity. The inactive space is commonly used to maintain a minimum pool and for sediment storage. Sediment storage may affect all levels of the reservoir storage. Also, the inactive capacity may sometimes be used during drought when it can provide limited but important storage for water supply, irrigation, recreation, fish and wildlife, and water quality.

e. Storage space allocation. Reservoir storage space may not be allocated to specific conservation purposes. Rather, reservoir releases can serve several purposes. However, the amount of water needed to serve each purpose varies. During drought, with limited multiple-purpose storage available, the purposes requiring greater releases begin to compete with purposes requiring less. For example, if the greater releases are not made, the storage would last longer for the purposes served by the lesser releases.

f. General information. A brief description of project purposes is presented below. Additional detail and a discussion of reservoir operating procedures may be found in EM 1110-2-3600, from which the following sections are excerpts.

2-3. Flood Control

a. Utilizing storage space. Reservoirs are designed to minimize downstream flooding by storing a portion or sometimes the entire runoff from minor or moderate flood events. Each reservoir's water control plan defines the goals of regulation. Usually, a compromise is achieved to best utilize the storage space to reduce flooding from both major and minor flood events. In special circumstances where reservoir inflows can be forecast several days or weeks in advance (for example, when the runoff occurs from snowmelt), for the best utilization of storage space, the degree of control for a particular flood event may be determined on the basis of forecasts. When runoff is seasonal, the amount of designated flood control storage space may be varied seasonally to better utilize the reservoirs for multiple-purpose regulation.

b. Releases. Flood control releases are based upon the overall objectives to limit the discharges at the downstream control points to predetermined damage levels. The regulation must consider the travel times caused by storage effects in the river system and the local inflows between the reservoir and downstream control points.

c. Intervening tributary and downstream damage areas. A multiple-reservoir system is generally regulated for flood control to provide flood protection both in intervening tributary areas and at downstream main stem damage areas. The extent of reservoir regulation required for protecting these areas depends on local conditions of flood damage, uncontrolled tributary drainage, reservoir storage capacity, and the volume and time distribution of reservoir inflows. Either the upstream or downstream requirements may govern the reservoir regulation, and usually the optimum regulation is based on the combination of the two.

d. Coordinated reservoir regulation. Water control with a system of reservoirs can incorporate the concept of a balanced reservoir regulation, with regard to filling the reservoirs in proportion to each reservoir's flood control capability, while also considering expected inflows and downstream channel capacities. Evacuation of flood water stored in a reservoir system must also be accomplished on a coordinated basis. Each reservoir in the system is drawn down as quickly as possible, considering conditions at control points, to provide space for controlling future floods. The objectives for withdrawal of water in the various zones of reservoir storage are determined to minimize the risk of encroaching into the flood control storage and to meet other project requirements. Sometimes the lower portion of the flood control pool must be evacuated slower to transition to a lower flow to minimize bank caving and allow channel recovery.

2-4. Navigation

a. Navigational requirements. Problems related to the management of water for navigation use vary widely among river basins and types of developments. Control structures at dams, or other facilities where navigation is one of the project purposes, must be regulated to provide required water flows and/or to maintain project navigation depths. Navigational requirements must be integrated with other water uses in multiple-purpose water resource systems. In the regulation of dams and reservoirs, the navigational requirements involve controlling water levels in the reservoirs and at downstream locations, and providing the quantity of water necessary for the operation of locks. There also may be navigational constraints in the regulation of dams and reservoirs with regard to rates of change of water surface elevations and outflows. There are numerous special navigational considerations that may involve water control, such as ice, undesirable currents and water flow patterns, emergency precautions, boating events, and launchings.

b. Waterflow requirements. Navigation locks located at dams on major rivers generally have sufficient water from instream flows to supply lockage water flow requirements. Navigation requirements for downstream use in open river channels may require larger quantities of water over a long period of time (from several months to a year), to maintain water levels for boat or barge transportation. Usually, water released from reservoirs for navigation is also used for other purposes, such as hydroelectric power, low-flow augmentation, water quality, enhancement of fish and wildlife, and recreation. Seasonal or annual water management plans are prepared which define the use of water for navigation. The amount of stored water to be released depends on the conditions of water storage in the reservoir system and downstream requirements or goals for low-flow augmentation, as well as factors related to all uses of the water in storage.

c. Using water for lockage. Navigational constraints are also important for short-term regulation of projects to meet all requirements. In some rivers, supply of water for lockage is a significant problem, particularly during periods of low flow or droughts. The use of water for lockage is generally given priority over hydropower or irrigation usages. However, this is dependent on the storage allocated to each purpose. In critical low-water periods, a curtailment of water use for lockage may be instituted by restricting the number of locks used, thereby conserving the utilization of water through a more efficient use of the navigation system. Water requirements for navigation canals are sometimes based on lockage and instream flows as necessary to preserve water quality in the canal.

2-5. Hydroelectric Power

a. Reservoir project categories. Reservoir projects which incorporate hydropower generally fall into two distinct categories: storage reservoirs which have sufficient capacity to regulate streamflow on a seasonal basis and run-of-river projects where storage capacity is minor relative to the volume of flow. Most storage projects are multiple-purpose. Normally, the upstream reservoirs include provisions for power production at the site, as well as for release of water for downstream control. Run-of-river hydropower plants are usually developed in connection with navigation projects.

b. Integration and control of a power system. Integration and control of a major power system involving hydropower resources is generally accomplished by a centralized power dispatching facility. This facility contains the equipment to monitor the entire power system operation, including individual plant generation, substation operation, transmission line operation, power loads and requirements by individual utilities and other bulk power users, and all factors related to the electrical system control for real-time operation. The dispatching center is manned on a continuous basis, and operations monitor and control the flow of power through the system, rectify outages, and perform all the necessary steps to ensure the continuity of power system operation in meeting system loads.

c. *Regulation of a hydropower system.* Regulation of hydropower systems involves two levels of control: scheduling and dispatching. The scheduling function is performed by schedulers who analyze daily requirements for meeting power loads and resources and all other project requirements. Schedules are prepared and thoroughly coordinated to meet water and power requirements of the system as a whole. Projections of system regulation, which indicate the expected physical operation of individual plants and the system as a whole, are prepared for one to five days in advance. These projections are updated on a daily or more frequent basis to reflect the continuously changing power and water requirements.

2-6. Irrigation

a. *Irrigation diversion requirements.* Irrigation water diverted from reservoirs, diversion dams, or natural river channels is controlled to meet the water duty requirements. The requirements vary seasonally, and in most irrigated areas in the western United States, the agricultural growing season begins in the spring months. The diversion requirements gradually increase as the summer progresses, reaching their maximum amounts in July or August. They then recede to relatively low amounts by late summer. By the end of the growing season, irrigation diversions are terminated, except for minor amounts of water that may be necessary for domestic use, stock water, or other purposes.

b. *Irrigation as project purpose.* Corps of Engineers' reservoir projects have been authorized and operated primarily for flood control, navigation, and hydroelectric power. However, several major Corps of Engineers multiple-purpose reservoir projects include irrigation as a project purpose. Usually, water for irrigation is supplied from reservoir storage to augment the natural streamflow as required to meet irrigation demands in downstream areas. In some cases, water is diverted from the reservoir by gravity through outlet facilities at the dam which feed directly into irrigation canals. At some of the run-of-river power or navigation projects, water is pumped directly from the reservoir for irrigation purposes.

c. *Meeting irrigation demands.* The general mode for regulation of reservoirs to meet irrigation demands is to capture all runoff in excess of minimum flow demands and water rights during the spring and early summer. This usually results in refilling the reservoirs prior to the irrigation demand season. The water is held in storage until the natural flow recedes to the point where it is no longer of sufficient quantity to meet all demands for downstream irrigation. At that time, the release of stored water from reservoirs is begun and continued on a demand basis until the end of the growing season (usually September or October). During the winter, projects release water as required for instream flows, stock water, or other project purposes.

2-7. Municipal and Industrial Water Supply

a. *Municipal and industrial use.* Regulation of reservoirs for municipal and industrial (M&I) water supply is performed in accordance with contractual arrangements. Storage rights of the user are defined in terms of acre-feet of stored water and/or the use of storage space between fixed limits of reservoir levels. The amount of storage space is adjusted to account for change in the total reservoir capacity that is caused by sediment deposits. The user has the right to withdraw water from the lake or to order releases to be made through the outlet works. This is subject to Federal restrictions with regard to overall regulation of the project and to the extent of available storage space.

b. *Temporary withdrawal.* In times of drought, special considerations may guide the regulation of projects with regard to water supply. Adequate authority to permit temporary withdrawal of water from Corps projects is contained in 31 U.S.C. 483a (HEC 1990e). Such withdrawal requires a fee that is sufficient to recapture lost project revenues, and a proportionate share of operation, maintenance, and major replacement expenses.

2-8. Water Quality

a. *Goal and objective.* Water quality encompasses the physical, chemical, and biological characteristics of water and the abiotic and biotic interrelationships. The quality of the water and the aquatic environment is significantly affected by management practices employed by the water control manager. Water quality control is an authorized purpose at many Corps of Engineers reservoirs. However, even if not an authorized project purpose, water quality is an integral consideration during all phases of a project's life, from planning through operation. The minimum goal is to meet State and Federal water quality standards in effect for the lakes and tailwaters. The operating objective is to maximize beneficial uses of the resources through enhancement and nondegradation of water quality.

b. Release requirements. Water quality releases for downstream control have both qualitative and quantitative requirements. The quality aspects relate to Corps' policy and objectives to meet state water quality standards, maintain present water quality where standards are exceeded, and maintain an acceptable tailwater habitat for aquatic life. The Corps has responsibility for the quality of water discharged from its projects. One of the most important measures of quality is quantity. At many projects authorized for water quality control, a minimum flow at some downstream control point is the primary water quality objective. Other common objectives include temperature, dissolved oxygen, and turbidity targets at downstream locations.

c. Coordinated regulation. Coordinated regulation of multiple reservoirs in a river basin is required to maximize benefits beyond those achievable with individual project regulation. System regulation for quantitative aspects, such as flood control and hydropower generation, is a widely accepted and established practice, and the same principle applies to water quality concerns. Water quality maintenance and enhancements may be possible through coordinated system regulation. This applies to all facets of quality from the readily visible quantity aspect to traditional concerns such as water temperature and dissolved oxygen content.

d. System regulation. System regulation for water quality is of most value during low-flow periods when available water must be used with greatest efficiency to avoid degrading lake or river quality. Seasonal water control plans are formulated based on current and forecasted basin hydrologic, meteorologic and quality conditions, reservoir status, quality objectives and knowledge of water quality characteristics of component parts of the system. Required flows and qualities are then apportioned to the individual projects, resulting in a quantitatively and qualitatively balanced system. Computer programs capable of simulating reservoir system regulation for water quality provide useful tools for deriving and evaluating water control alternatives.

2-9. Fish and Wildlife

Project regulation can influence fisheries both in the reservoir pool and downstream. One of the most readily observable influences of reservoir regulation is reservoir pool fluctuation. Periodic fluctuations in reservoir water levels present both problems and opportunities to the water control manager with regard to fishery management. The seasonal fluctuation that occurs at many flood control reservoirs, and the daily fluctuations that occur with hydropower operation often result in the elimination of shoreline vegetation and subsequent shoreline erosion, water quality degradation and loss of habitat. Adverse impacts of water level fluctuations also include loss of shoreline shelter and physical disruption of spawning and nests.

2-10. Recreation

a. Reservoir level. Recreational use of the reservoirs may extend throughout the entire year. Under most circumstances, the optimum recreational use of reservoirs would require that the reservoir levels be at or near full conservation pool during the recreation season. The degree to which this objective can be met varies widely, depending upon the regional characteristics of water supply, runoff, and the basic objectives of water regulation for the various project purposes. Facilities constructed to enhance the recreational use of reservoirs may be designed to be operable under the planned reservoir regulation guide curves on water control diagrams, which reflect the ranges of reservoir levels that are to be expected during the recreational season.

b. Downstream river levels. In addition to the seasonal regulation of reservoir levels for recreation, regulation of project outflows may encompass requirements for specific regulation criteria to enhance the use of the rivers downstream from the projects, as well as to ensure the safety of the general public. The Corps has the responsibility to regulate projects in a manner to maintain or enhance the recreational use of the rivers below projects to the extent possible (i.e., without significantly affecting the project function for authorized purposes). During the peak recreation season, streamflows are regulated to ensure the safety of the public who may be engaged in water related activities, including boating, swimming, fishing, rafting, and river drifting. Also, the aesthetics of the rivers may be enhanced by augmenting streamflows during the low-water period. Water requirements for maintaining or enhancing the recreational use of rivers are usually much smaller than other major project functional uses. Nevertheless, it is desirable to include specific goals to enhance recreation in downstream rivers in the water control plan. The goals may be minimum project outflows or augmented streamflows at times of special need for boating or fishing. Of special importance is minimizing any danger that might result from changing conditions of outflows which would cause unexpected rise or fall in river levels. Also, river drifting is becoming an important recreational use of rivers, and in some cases it may be possible to enhance the conditions of stream flow for relatively short periods of time for this purpose.

2-11. Water Management Goals and Objectives

a. Water management. ER 1110-2-240 paragraph 6, defines the goals and objectives for water regulation by the Corps. Basically, the objective is to conform with specific provisions of the project authorizing legislation and water management criteria defined in Corps of Engineers reports prepared in the planning and design of the project or system. Beyond this, the goals for water management will include the provisions, as set forth in any applicable authorities, established project construction, and all applicable Congressional Acts and Executive Orders relating to operations of Federal facilities.

b. Water control systems management. EM 1110-2-3600 provides guidance on water control plans and project management. A general prime requirement in project regulation is the safety of users of the facilities and the general public, both at the project and at downstream locations. The development of water control plans and the scheduling of reservoir releases must be coordinated with appropriate agencies, or entities, as necessary to meet commitments made during the planning and design of the project. Additionally, water control plans must be reviewed and adjusted, when possible, to meet changing local conditions.

c. Regional management. Regional water management should consider the interaction of surface-ground-water resources. HEC Research Document 32 provides examples for several regions in the United States (HEC 1991c).

Chapter 3
Multiple-Purpose Reservoirs

3-1. Hydrologic Studies for Multipurpose Projects

a. Conception. Multipurpose reservoirs were originally conceived as projects that served more than one purpose independently and would effect savings through the construction of a single large project instead of two or more smaller projects. As the concept developed, the joint use of water and reservoir space were added as multipurpose concepts. Even such competitive uses as flood control and water supply could use the same reservoir space at different times during the year.

b. Feasibility. The feasibility of multiple-purpose development is almost wholly dependent upon the demonstrated ability of a proposed project to serve several purposes simultaneously without creating conditions that would be undesirable or intolerable for the other purposes. In order to demonstrate that multipurpose operation is feasible, detailed analyses of the effects of various combinations of streamflows, storage levels, and water requirements are required. Detailed analyses of these factors may be overlooked during the planning phase because the analyses are complex and simplifying methods or assumptions may not consider some details that may be important. However, ignoring the details of multipurpose operation in the planning phase is risky because the operation criteria are critical in determining the feasibility of serving several purposes simultaneously.

c. Defining the multipurpose project. One of the factors that make detailed sequential analyses of multipurpose operation difficult during planning studies is that sufficient data on various water demands are either not available or not of comparable quality for all purposes. To adequately define the multipurpose operation, the analyses must include information on the magnitude and seasonal variations of each demand, long-term changes in demands, relative priority of each use, and shortage tolerances. Information on magnitude and seasonal variation in demands and on long-term variations in demands is usually more readily available than information on relative priorities among uses and on shortage tolerances. If information on priorities and shortages is not available from the various users, one can make several assumptions concerning the priorities and perform sequential routing studies for each set of assumptions. The results of these studies can determine the consequences of various priorities to potential water users. It may be possible for the potential users to adopt a priority arrangement based on the value of the water for the various demands.

d. Success of multipurpose projects. The success of multipurpose operation also depends on the formulation of operational rules that ensure that water in the proper quantities and qualities is available for each of the purposes at the proper time and place. Techniques for formulating operational rules are not fixed, but the logical approach involves determining the seasonal variation of the flood-control space requirement, and the seasonal variation of conservation requirements, formulation of general operational rules that satisfy these requirements, and detailed testing of the operational rules to ascertain the adequacy of the plan for each specific purpose.

e. Multipurpose project rules. The judgment of an experienced hydrologic engineer is invaluable in the initial formulation and subsequent development and testing of operational rules. Although the necessary rules cannot be completely developed until most of the physical dimensions of the project are known, any tendency to discount the importance of operational rules as a planning variable should be resisted because of the important role they often assume in the feasibility of multipurpose projects. As a minimum, the operational rules used in a planning study should be sufficiently refined to assist the engineer in evaluating the suitability of alternative projects to satisfy water demands for specified purposes.

3-2. Relative Priorities of Project Functions

a. Developing project rules. As indicated above, the use of operational rules based on the relative priorities among the project purposes appears to offer the best approach to multipurpose operational problems. The degree of success that can be realized depends on a realistic priority system that accurately reflects the relative value of water from the project for a given purpose at a given time. Unless a realistic priority system is used to develop the operational rules, it will not be possible to follow the rules during the project life because the true priorities may control the operational decisions and prevent the project from supplying the services it was designed to provide.

b. Typical system priorities. Priorities among the various water resource purposes vary with locale, water rights, the need for various types of water use, the legal and political considerations, and with social, cultural, and environmental conditions. Although these variations make it impossible to specify a general priority system, it is useful to identify a set of priorities that would be typical under average conditions. In such a situation, operation for the safety of the structure has the highest priority unless the consequences of failure of the structure are minor (which is seldom the case). Of the functional purposes, flood control must have a high priority, particularly where downstream levees, bridges, or other vital structures are threatened. It is not unusual for conservation operations to cease entirely during periods of flood activity if a significant reduction in flooding can be realized thereby. Among the conservation purposes, municipal and industrial water supply and hydroelectric power generation are often given a high priority, particularly where alternative supplies are not readily available. After those purposes, other project purposes usually have a somewhat lower priority because temporary shortages are usually not disastrous. It should be emphasized again that there can be marked exceptions to these relative priorities. There are regional differences in relative needs and, legal and institutional factors may greatly affect priorities.

c.   Complex system priorities. In complex reservoir systems, with competing demands and several alternative projects to meet the demand, the relative priority among projects and purposes may not be obvious. The operation rules, which can be evaluated with detailed simulation, may not be known or may be subject to criticism. In these situations, it may be useful to apply a system analysis based on consistent values for the various project purposes. The results of the analysis could suggest an operational strategy which can be tested with more detailed analysis. Chapter 4 presents information and approaches for system analysis.

3-3. Managing Competitive and Complementary Functions

a.   Identifying interactions between purposes. Before operation rules can be formulated, the adverse (competitive) and the beneficial (complementary) interactions between purposes must be identified. The time of occurrence of the interactions is often as important as the degree of interaction, particularly if one or more of the water uses has significant variations in water demand. In supplying water from a single reservoir for several purposes with seasonally varying demands, it is possible for normally complementary purposes to become competitive at times due to differences in their seasonal requirements.

b.   Allocating storage space. When several purposes are to be served from a single reservoir, it is possible to allocate space within certain regions of the reservoir storage for each of the purposes. This practice evolved from projects that served only flood control and one conservation purpose because it was necessary to reserve a portion of the reservoir storage for storing floodwater. It is still necessary to have a specific allocation of flood-control storage space (although the storage reservation can vary seasonally) because of the basic conflict between reserving empty storage space for regulating potential floods and filling that space to meet future water supply requirements. However, applying specific storage allocations or reservations for competing conservation purposes should be kept to a minimum because it reduces operational flexibility.

c.   Operational conflicts. Allocation of specific storage space to several purposes within the conservation pool can result in operational conflicts that might make it impossible or very costly to provide water for the various purposes in the quantities and at the time they are needed. The concept of commingled or joint-use conservation storage for all conservation purposes with operational criteria to maximize the complementary effects and minimize the competitive effects is far easier to manage and, if carefully designed, will provide better service for all purposes. Where the concept of joint-use storage is used, the operational criteria should be studied in the planning process in such a way that the relative priorities of the various purposes are taken into account. This allows careful evaluation of a number of priority systems and operational plans. The operational decisions that result from such disputes are frequently not studied in enough detail (from the engineering point of view), and as a result, the ability of the project to serve some purposes may be seriously affected.

3-4. Operating Concepts

a.   Operating goals. Reservoir operating goals vary with the storage in the reservoir. The highest zone in the reservoir is that space reserved at any particular time for the control of floods. This zone includes the operational flood-control space and the surcharge space required for the passage of spillway flows. Whenever water is in this zone it must be released in accordance with flood-control requirements. The remaining space can be designated as conservation space. The top zone of conservation space may include storage that is not required to satisfy the firm conservation demands, including recreational use of the reservoir. Water in this space can be released as surplus to serve needs or uses that exceed basic requirements. The middle zone of conservation space is that needed to store water to supply firm water needs. The bottom zone of conservation space can be termed buffer space, and when operation is in this zone the firm services are curtailed in order to prevent a more severe shortage later. The bottom zone of space in the reservoir is designated as the minimum pool reserved for recreation, fish, minimum power head, sediment reserve, and other storage functions.

b. Storage zone boundaries. The boundaries between storage zones may be fixed at a constant level or they may vary seasonally. In general, the seasonally varying boundaries offer the potential for a more flexible operating plan that can result in higher yields for all purposes. However, the proper location of the seasonal boundaries requires more study than the location of a constant boundary. This is discussed in more detail in Chapter 11. Furthermore, an additional element of chance is introduced when the boundaries are allowed to vary, because the joint use of storage might endanger firm supplies for one or more specific purposes. The location of the seasonally varying boundaries is determined by a process of formulating a set of boundaries and attendant operational rules, testing the scheme by a detailed sequential routing study, evaluating the outcome of the study, changing the rules or boundaries if necessary, and repeating the procedure until a satisfactory operation results.

c. Demand schedules. Expressing demand schedules as a function of the relative availability of water is another means of incorporating flexibility and relative priority in operational rules. For example, the balance between hydro and thermal power generation might well be a continuous function of available storage. As another example, it might be possible to have two or more levels of navigation service or lengths of navigation season with the actual level of service or length of season being dependent upon the availability of water in the reservoir. By regulating the level of supply to the available water in the reservoir, users can plan emergency measures that will enable them to withstand partial reductions in service and thereby avoid complete cessation of service, which might be disastrous. Terms such as desired flow and minimum required flow for navigation can be used to describe two levels of service.

d. Levels of service. There can be as many levels of service as a user desires, but each level requires criteria for determining when the level is to be initiated and when it is to be terminated. The testing and development of the criteria for operating a multipurpose project with several purposes and several levels of service are accomplished by detailed sequential routing studies. Because the development and testing of these criteria are relatively difficult, the number of levels of service should be limited to the minimum number needed to achieve a satisfactory plan of operation.

e. Buffer storage. Buffer storage or buffer zones are regions within the conservation storage where operational rules effect a temporary reduction in firm services. The two primary reasons for temporarily reducing services are to ensure service for a high-priority purpose while eliminating or curtailing services for lower-priority purposes, and to change from one level of service for a given purpose to a lower level of service for that same purpose when storage levels are too low to ensure the continuation of firm supplies for all purposes. As with the other techniques for implementing a multipurpose operation, the amount of buffer storage and the location of the boundaries cannot be determined accurately except by successive approximations and testing by sequential routing studies.

3-5. Construction and Physical Operation

a. General. In addition to hydrologic determinations discussed above, a number of important hydrologic determinations are required during project construction and during project operation for ensuring the integrity of the project and its operation.

b. Cofferdams. From a hydrologic standpoint, during construction the provisions for streamflow diversion are a primary concern. If a cofferdam used for dewatering the work area is overtopped, serious delays and additional construction costs can result. In the case of high cofferdams where substantial poundage occurs, it is possible that failure could cause major damage in downstream areas. Cofferdams should be designed on the same principles as are permanent dams, generally on the basis of balancing incremental costs against incremental benefits of all types. This will require flood frequency and hypothetical flood studies, as described in Chapters 6 and 7 of this manual. Where major damage might result from cofferdam failure, a standard project flood (SPF) or even a probable maximum flood (PMF) may be used as a primary basis for design.

c. Overtopping. Where a major dam embankment may be subject to overtopping during construction, the diversion conduit capacity must be sufficient to regulate floods that might occur with substantial probability during the critical construction period. It is not necessary that the regulated releases be nondamaging downstream, but it is vital that the structure remain intact. An explicit evaluation of risk of embankment failure and downstream impacts during construction should be presented in the design document.

d. Conduits, spillways, and gates. Conduits, spillways, and all regulating gates must be functionally adequate to accomplish project objectives. Their sizes, dependability, and speed of operation should be tested using recorded and hypothetical hydrographs and anticipated hydraulic heads to ensure that they will perform properly. The nature of stilling facilities might be dictated by hydrologic considerations if frequency and duration of high outflows substantially influence their design. The necessity for multilevel intakes to control the quality of reservoir releases can be assessed by detailed reservoir stratification studies under all combinations of hydrologic and reservoir conditions. Techniques for conducting reservoir stratification studies are discussed in EM 1110-2-1201.

e. Design. The design of power facilities can be greatly influenced by hydrologic considerations, as discussed in Chapter 11 of this manual and EM 1110-2-1701. General considerations in the hydrologic design of spillways are discussed in Chapter 10 and more detailed information is presented in Chapter 14 herein.

f. Extreme floods. Regardless of the reservoir purposes, it is imperative that spillway facilities provided will ensure the integrity of the project in the event of extreme floods. Whenever the operation rules of a reservoir are substantially changed, spillway facilities should be reviewed to ensure that the change in project operation does not adversely alter the capability to pass extreme floods without endangering the structure. The capability of a spillway to pass extreme floods can be adversely affected by changes in operation rules that actually affect the flood operation itself or by changes that result in higher pool stages during periods of high flood potential.

g. Special operating rules. A number of situations might require special operating rules. For example, operating rules are needed for the period during which a reservoir is initially filling, for emergency dewatering of a reservoir, for interim operation of one or more components in a system during the period while other components are under construction, and for unanticipated conditions that seem to require deviation from established operating rules. The need for operation rules during the filling period is especially important because many decisions must be based on the filling plan. Among the important factors that are dependent upon the filling schedule are the on-line date for power generating units, the in-service dates for various purposes such as water supply and navigation, and the effective date for legal obligations such as recreation concessions.

h. Specification of monitoring facilities. One of the more important considerations in the hydrologic analysis of any reservoir is the specification of monitoring facilities, including streamflow, rainfall, reservoir stage, and other hydrologic measurements. These facilities serve two basic purposes: to record all operations and to provide information for operation decisions. The former purpose satisfies legal requirements and provides data for future studies. The latter purpose may greatly increase the project

effectiveness by enabling the operating agency, through reliable forecasts of hydrologic conditions, to increase operation efficiency. Hydrologic aspects of monitoring facilities and forecasts will be presented in a new EM on hydrologic forecasting.

i. Stream gauges. Because gauged data are most important during flood events, special care should be taken in locating the gauge. Stream gauges should not be located on bridges or other structures that are subject to being washed out. To the extent possible, the gauges should be capable of working up through extreme flood events, and stage-discharge relationships should be developed up to that level. The gauge should have reasonable access for checking and repair during the flood. Reservoir spilling, local flooding, and backwater effects from downstream tributaries should all be considered when finding a suitable location. More detailed information on stream gauges can be found in many USGS publications, such as Carter and Davidian (1968), Buchanan and Somers (1968 and 1969), or Smoot and Novak (1969).

3-6. General Study Procedure

As indicated earlier, there is no fixed procedure for developing reservoir operational plans for multipurpose projects; however, the general approach that should be common to all cases would include the following steps:

a. Survey the potential water uses to be served by the project in order to determine the magnitude of each demand and the seasonal and long-term variations in the demand schedule.

b. Develop a relative priority for each purpose and determine the levels of service and required priority that will be necessary to serve each purpose. If necessary, make sequential studies illustrating the consequences of various alternative priority systems.

c. Establish the seasonal variation of flood-control space required, using procedures discussed in Chapter 10.

d. Establish the total power, water supply, and low-flow regulation requirements for competitive purposes during each season of the year.

e. Establish preliminary feasibility of the project based on physical constraints.

f. Establish the seasonal variation of the storage requirement to satisfy these needs, using procedures described in Chapter 11.

g. Determine the amount of storage needed as a minimum pool for power head, recreation, sedimentation reserve, and other purposes.

h. Using the above information, estimate the size of reservoir and seasonal distribution of space for the various purposes that would satisfy the needs. Determine the reservoir characteristics, including flowage, spillway, power plant, and outlet requirements.

i. Test and evaluate the operation of the project through the use of recorded hydrologic data in a sequential routing study to determine the adequacy of the storage estimates and proposed rules with respect to the operational objectives for each purpose. If the record is short, supplement it with synthetic floods to evaluate flood storage reserves. If necessary, make necessary changes and repeat testing, evaluating, and changing until satisfactory operation is obtained.

j. Test proposed rules of operation by using sequential routing studies with stochastic hydrologic data to evaluate the possibility of historical bias in the proposed rules.

k. Determine the needs for operating and monitoring equipment required to ensure proper functional operation of the project.

l. As detailed construction plans progress, evaluate cofferdam needs and protective measures needed for the integrity of project construction, particularly diversion capacity as a function of dam construction stage and flood threat for each season.

Chapter 4
Reservoir Systems

4-1. Introduction

Water resource systems should be designed and operated for the most effective and efficient accomplishment of overall objectives. The system usually consists of reservoirs, power plants, diversions, and canals that are each constructed for specific objectives and operated based on existing agreements and customs. Nevertheless, there is considerable latitude in developing an operational plan for any water resource system, but the problem is greatly complicated by the legal and social restrictions that ordinarily exist.

a. Mathematical modeling. Water resource system operation is usually modeled mathematically, rather than with physical models. The mathematical representation of a water resource system can be extremely complex. Operations research techniques such as linear programming and dynamic programming can be applied to a water resource system; however, they usually are not capable of incorporating all the details that affect system outputs. It is usually necessary to simulate the detailed sequential operation of a system, representing the manner in which each element in the system will function under realistic conditions of inputs and requirements on the system. The simulation can be based on the results from the optimization of system outputs or repeated simulations. Successively refining the physical characteristics and operational rules can be applied to find the optimum output.

b. Inputs and requirements. A factor that greatly complicates the simulation and evaluation of reservoir system outputs is the stochastic nature of the inputs and of the requirements on the system. In the past, it has been customary to evaluate system accomplishments on the assumption that a repetition of historical inputs and requirements (adjusted to future conditions) would adequately represent system values. However, this assumption has been demonstrated to be somewhat deficient. It is desirable to test any proposed system operation under a great many sequences of inputs and requirements. This requires a mathematical model that will define the frequency and correlation characteristics of inputs and requirements and that is capable of generating a number of long sequences of these quantities. Concepts for accomplishing this are discussed in paragraph 5-5.

4-2. System Description

a. Simulating system operation. Water resource systems consist of reservoirs, power plants, diversion structures, channels, and conveyance facilities. In order to simulate system operation, the system must be completely described in terms of the location and functional characteristics of each facility. The system should include all components that affect the project operation and provide the required outputs for analysis.

(1) Reservoirs. For reservoirs, the relation of surface area and release capacity to storage content must be described. Characteristics of the control gates on the outlets and spillway must be known in order to determine constraints on operation. The top-of-dam elevation must be specified and the ability of the structure to withstand overtopping must be assessed.

(2) Downstream channels. The downstream channels must be defined. Maximum and minimum flow targets are required. For short-interval simulation, the translation of flow through the channel system is modeled by routing criteria. The travel time for flood flow is important in determining reservoir releases and potential limits for flood control operation to distant downstream locations.

(3) Power plants. For power plants at storage reservoirs, the relation of turbine and generation capacity to head must be determined. To compute the head on the plant, the relation of tailwater elevation to outflow must be known. Also, the relation of overall power plant efficiency to head is required. Other characteristics such as turbine leakage and operating efficiency under partial load are also important.

(4) Diversion structures. For diversion structures, maximum diversion and delivery capacity must be established. The demand schedule is required, and the consumptive use and potential return flow to the system may be important for the simulation.

b. Preparing data. While reservoir system data must be defined in sufficient detail to simulate the essence of the physical system, preparing the required hydrologic data may require far more time and effort. The essential flow data are required for the period of record, for major flood events, and in a consistent physical state of the system. Flow records are usually incomplete, new reservoirs in the system change the flow distribution, and water usage in the watershed alters the basin yield over time. Developing a consistent hydrologic data series, making maximum use of the available information, is discussed in Chapter 5.

4-3. Operating Objectives and Criteria

a. User services. Usually, there is a fixed objective for each function in a water resource system. Projects are constructed and operated to provide services that are counted on by the users. In the case of power generation and water supply, the services are usually contracted, and it is essential to provide contracted amounts insofar as possible. Services above the contracted amounts are ordinarily of significantly less value. Some services, such as flood control and recreation, are not ordinarily covered by contracts. For these, service areas are developed to provide the degree of service for which the project was constructed.

b. Rules for services. Shortages in many of the services can be very costly, whereas surpluses are usually of minor value. Accordingly, the objectives of water resource system operational are usually fixed for any particular plan of development. These are expressed in terms of operational rules that specify quantities of water to be released and diverted, quantities of power to be generated, reservoir storage to be maintained, and flood releases to be made. These quantities will normally vary seasonally and with the amount of storage water in the system. Rule curves for the operation of the system for each function are developed by successive approximations on the basis of performance during a repetition of historical streamflows, adjusted to future conditions, or on the basis of synthetic stream flows that would represent future runoff potential.

4-4. System Simulation

The evaluation of system operation under specified operation rules and a set of input quantities is complex and requires detailed simulation of the operation for long periods of time. This is accomplished by assuming that steady-state conditions prevail for successive intervals of time. The time interval must be short enough to capture the details that affect system outputs. For example, average monthly flows may be used for most conservation purposes; however, for small reservoirs, the flow variation within a month may be important. For hydropower reservoirs, the average monthly pool level or tailwater elevation may not give an accurate estimate of energy production.

To simulate the operation during each interval, the simulation solves the continuity equation with the reservoir release as the decision variable. The system is analyzed in an upstream-to-downstream direction. At each pertinent location, requirements for each service are noted, and the reservoirs at and above that location are operated in such a way as to serve those requirements, subject to system constraints such as outlet capacity, and channel capacity, and reservoir storage capacity. As the computation procedure progresses to downstream locations, the tentative release decisions made for upstream locations become increasingly constraining. It often becomes necessary to assign priorities among services that conflict. Where power generation causes flows downstream to exceed channel capacity, for example, a determination must be made as to whether to curtail power generation. If there is inadequate water at a diversion to serve both the canal and river requirements, a decision must be made.

4-5. Flood-Control Simulation

Flood discharge can change rapidly with time. Therefore, steady-state conditions cannot be assumed to prevail for long periods of time (such as one month). Also, physical constraints such as outlet capacity and the ability to change gate settings are more important. The time translation for flow and channel storage effects cannot ordinarily be ignored. Consequently, the problem of simulating the flood-control operation of a system can be more complex than for conservation.

a. Computational interval. The computation interval necessary for satisfactory simulation of flood operations is usually on the order of a few hours to one day at the most. Sometimes intervals as short as 15 or 30 min are necessary. It is usually not feasible to simulate for long periods of time, such as the entire period of record, using such a short computation interval. However, period-of-record may be unnecessary because most of the flows are of no consequence from a flood-control standpoint. Accordingly, simulation of flood-control operation is usually made only for important flood periods.

b. Starting conditions. The starting conditions for simulating the flood-control operation for an historic flood period would depend on the operation of the system for conservation purposes prior to that time. Accordingly, the conservation operation could be simulated first to establish the state of the system at the beginning of the month during which the flood occurred as the initial conditions for the flood simulation. However, the starting storage for flood operation should be based on a realistic assessment of likely future conditions. If it is likely that the conservation pool is full when a flood occurs, then that would be a better starting condition to test the flood-pool capacity. It may be possible that the starting pool would be higher if there were several storms in sequence, or if the flood operation does not start the instant excessive inflows raise the pool level into flood-control space.

c. *Historic sequences.* While simulating historic sequences are important, future floods will be different and occur in different sequences. Therefore, the analysis of flood operations should utilize both historic and synthetic floods. The possibility of multiple storms, changes in the upstream catchment, and realistic flood operation should be included in the analysis. Chapter 7 presents flood-runoff analysis and Chapter 10 presents flood-control storage requirements.

d. *Upstream-to-downstream solution.* If the operation of each reservoir in a system can be based on conditions at or above that reservoir, an upstream-to-downstream solution approach can establish reservoir releases, and these releases can be routed through channel reaches as necessary in order to obtain a realistic simulation. Under such conditions, a simple simulation model is capable of simulating the system operation with a high degree of accuracy. However, as the number of reservoirs and downstream damage centers increase, the solution becomes far more complex. A priority criteria must be established among the reservoirs to establish which should release water, when there is a choice among them.

e. *Combination releases.* The HEC-5 *Simulation of Flood Control and Conservation Systems* (HEC 1982c) computer program can solve for the combination of releases at upstream reservoirs that will satisfy channel capacity constraints at a downstream control point, taking into account the time translation and channel storage effects, and that will provide continuity in successive time intervals. The time translation effects can be modeled with a choice of hydrologic routing methods. Reservoir releases are determined for all designated downstream locations, subject to operation constraints. The simulation is usually performed with a limited foresight of inflows and a contingency factor to reflect uncertainty in future flow values. The concept of pool levels is used to establish priorities among projects in multiple-reservoir systems. Standard output includes an indicator for the basis of reservoir release determination, along with standard simulation output of reservoir storage, releases, and downstream flows.

f. *Period-of-record flows.* Alternatively, a single time interval, such as daily, can be used to simulate period-of-record flows for all project purposes. This approach is routinely used in the Southwestern Division with the computer program "Super" (USACE 1972), and in the North Pacific Division with the SSARR program (USACE 1991). The SSARR program is capable of simulation on variable time intervals.

4-6. Conservation Simulation

While the flood-control operation of a reservoir system is sensitive to short time variations in system input, the operation of a system for most conservation purposes is usually sensitive only to long-period streamflow variations. Historically, simulation of the conservation operation of a water resource system has been based on a relatively long computation interval such as a month. With the ease of computer simulation and available data, shorter computational intervals (e.g., daily) can provide a more accurate accounting of flow and storage. Some aspects of the conservation operation, such as diurnal variations in power generation in a peaking project, might require even shorter computational intervals for selected typical or critical periods to define important short-term variations.

a. *Hydropower simulation.* Hydropower simulation requires a realistic estimate of power head, which depends on reservoir pool level, tailwater elevation, and hydraulic energy losses. Depending on the size and type of reservoir, there can be considerable variation in these variables. Generally, the shorter time intervals will provide a more accurate estimate of power capacity and energy productions.

b. *Evaporation and channel losses.* In simulating the operation of a reservoir system for conservation, the time of travel of water between points in the system is usually ignored, because it is small in relation to the typical computation interval (e.g., monthly or weekly). On the other hand, evaporation and channel losses might be quite important; and it is sometimes necessary to account for such losses in natural river channels and diversion canals.

c. *Rule curves.* Rule curves for the operation of a reservoir system for conservation usually consist of standard power generation and water supply requirements that will be served under normal conditions, a set of storage levels that will provide a target for balancing storage among the various system reservoirs, and maximum and minimum permissible pool levels for each season based on flood control, recreation, and other project requirements. Often some criteria for decreasing services when the system reservoir storage is critically low will be desirable.

4-7. System Power Simulation

Where a number of power plants in the water resource system serve the same system load, there is usually considerable flexibility in the selection of plants for power generation at any particular time. In order to simulate the operation of the system for power generation, it is necessary to specify the overall system requirement and the minimum amount of energy that must be generated at each plant during each month or other interval of time. Because the entire system power requirement might possibly be supplied by incidental generation due to releases made for other purposes, it is first necessary to search the entire system to determine generation that would occur with only minimum power requirements at each plant and with all requirements throughout the system for other purposes. If insufficient power is generated to meet the entire system load in this manner, a search will be made for those power reservoirs where storage is at a higher level, in relation to the rule curves, than at other power reservoirs. The additional power load requirement will then be assigned to those reservoirs in such a manner as to maintain the reservoir storage as nearly as possible in conformance with the rule curves that balance storage among the reservoirs in the most desirable way. This must be done without assigning more power to any plant than it can generate at overload capacity and at the system load factor for that interval. EM 1110-2-1701 paragraph 5-14, describes hydropower system analysis.

4-8. Determination of Firm Yield

If the yield is defined as the supply that can be maintained throughout the simulation period without shortages, then the process of computing the maximum yield can be expedited. This is done by maintaining a record of the minimum reserve storage (if no shortage has yet occurred) or of the amount of shortage (if one does occur) in relation to the total requirement since the last time that all reservoirs were full. The surplus or shortage that existed at the end of any computation interval would be expressed as a ratio of the supply since the reservoirs were last full, and the minimum surplus ratio (if no shortage occurs) or maximum shortage ratio (if a shortage does occur) that occurs during the entire simulation period would be used to adjust the target yield for the next iteration. This basic procedure for computing firm yield is included in the HEC-5 computer program. Additionally, the program has a routine to make an initial estimate of the critical period and expected yield. After the yield is determined using the critical period, the program will evaluate the yield by performing a simulation with the entire input flow record. Chapter 12 describes storage-yield procedures.

4-9. Derivation of Operating Criteria

A plan of development for a water resource system consists not only of the physical structures and their functional characteristics but also of the criteria by which the system will be operated. In order to compare alternative plans of development, it is necessary that each plan be operated optimally. The derivation of optimal operation criteria for a water resource system is probably more difficult than the derivation of optimum configuration and unit sizes because any small change in operation rules can affect many functions in the system for long periods of time and in very subtle ways.

a. Simulation. Operation criteria generally consist of release schedules at reservoirs, diversion schedules at control points, and minimum flows in the river at control points, in conjunction with reservoir balancing levels that define the target storage contribution among the various reservoirs in the system. All of these can vary seasonally, and target flows can vary stochastically. Once the unit sizes and target flows are established for a particular plan of development, a system of balancing levels must be developed. The system response to a change in these balancing levels is a complicated function of many system, input, and requirement characteristics. For this reason, the development of a set of balancing levels is an iteration process, and a complete system simulation must be done for each iteration.

(1) When first establishing balancing levels in the reservoir system, it usually is best to simulate system operation only for the most critical periods of historical streamflows. The final solution should be checked by simulation for long periods of time. The balancing levels defining the flood-control space are first tentatively established on the basis of minimum requirements for flood control storage that will provide the desired degree of protection. Preliminary estimates of other levels can be established on the basis of reserving the most storage in the smaller reservoirs, in those reservoirs with the least amount of runoff, and in those reservoirs that supply operation services not producible by other reservoirs.

(2) After a preliminary set of balancing levels is established, they should be defined approximately in terms of a minimum number of variables. The general shape and spacing of levels at a typical reservoir might be defined by the use of four or five variables, along with rules for computing the levels from those variables. Variations in levels among reservoirs should be defined by one or two variables, if possible, in order to reduce the amount of work required for optimization to an acceptable quantity.

(3) Optimization of a set of balancing levels for operational rule curves can be accomplished by successive approximations using a complete system simulation computation for critical drought periods. However, the procedures are limited to the input specifications of demands and storage allocation. While one can compare simulation results and conclude one is better than another based on

performance criteria, there is no way of knowing that an optimum solution has been achieved.

b. Optimization. While water resource agencies have generally focused on simulation models for system analysis, the academic community and research literature have emphasized optimization and stochastic analysis techniques. Research performed at HEC (HEC 1991b) has found a proliferation of papers on optimization of reservoir system operations written during the past 25 years, primarily by university researchers. There still remains a considerable gap between the innovative applications reported in the literature and the practices followed by the agencies responsible for water resource development. One basic problem is that many of the reported applications are uniquely formulated to solve a specific problem for a given system. There is a general view that the models performance, or the methods assumptions, would not sufficiently evaluate a different problem and system.

c. Prescriptive reservoir model. HEC has developed a system analysis tool based on a network flow model (HEC 1991a). The Prescriptive Reservoir Model (HEC-PRM) will identify the water allocation that minimizes poor performance for all defined system purposes. Performance is measured with analyst-provided functions of flow or storage or both. The physical system is represented as a network, and the allocation problem is formulated as a minimum-cost network flow problem. The objective functions for this network problem are convex, piecewise-linear approximations of the summed penalty functions for each project purpose (HEC 1991d).

(1) Systems have been analyzed in studies on the Missouri River (HEC 1991d) and the Columbia River (HEC 1991f). A preliminary analysis of the Phase I Missouri River study has developed initial methodologies for developing operation plans based on PRM results (HEC 1992b). Continued application experience is required to define generalized procedures for these analyses.

(2) The primary advantages for the HEC-PRM approach are the open state of the system and the required penalty functions for each system purpose. There are no rule curves or details of storage allocation, only basic physical constraints are defined. The reservoir system information defines maximum and minimum storage in the reservoirs and the linking of the system through the network of channels and diversions. The other primary reservoir data is traditional period-of-record monthly flows for the system.

(3) The development of the penalty functions requires an economic evaluation of the values to be placed on flow and storage in the system. The process is difficult and there are disagreements on the values, due to the difficulty of defining values for some purposes. However, the process does provide a method for defining and reviewing the purposes and their relative values.

(4) The primary disadvantage of the HEC-PRM is that the monthly flow data and lack of channel routing limit its application for short interval simulation, such as flood control and peaking hydropower. Additionally, the optimized solution is provided in terms of period-of-record flows and storage; however, the basis for the system operation are not explicitly defined. The post-processing of the results requires interpretation of the results in order to develop an operation plan that could be used in basic simulation and applied operation. More experience with this analysis of results is still required to define these procedures.

4-10. System Formulation Strategies

a. Determining the best system. A system is best for the national income criteria if it results in a value for system net benefits that exceeds that of any other feasible system. Except where noted, the following discussion was developed in a paper presented at the International Commissions on Large Dams Congress (Eichert and Davis 1976). For a few components, analysis of the number of alternative systems that are feasible is generally manageable, and exhaustive evaluation provides the strategy for determining the best system. When the number of components is more than just a few, then the exhaustive evaluation of all feasible alternative systems cannot practically be accomplished. In this instance, a strategy is needed that reduces the number of system alternatives to be evaluated to a manageable number while providing a good chance of identifying the best system. System analysis does not permit (maximum net benefit system) for reasonably complex systems even with all hydrologic-economic data known. An acceptable strategy need not make the absolute guarantee of economic optimum because seldom will the optimum economic system be selected as best.

b. Incremental test. The incremental test of the value of an individual system component is definitive for the economic efficiency criteria and provides the basis for several alternative formulation strategies. If existing reservoir components are present in the system, then they define the base conditions. If no reservoirs exist, the base condition would be for natural conditions. The strategies described below are extensions of currently used techniques and are based upon the concept of examining in detail the performance of a selected few alternative systems. The performance is assumed to be evaluated generally by traditional simulation methods, like the use of HEC-5.

c. Reasoned thought strategy. This strategy is predicated upon the idea that it is possible to reason out using judgment and other criteria, reasonable alternative systems. The strategy consists of devising through rational thought, sampling, public opinion, literature search, and brainstorming, a manageable number of system alternatives that will be evaluated. No more than 15 to 20 alternative systems could be evaluated by detailed simulation in a practical sense.

(1) The total performance of each system in terms of economic (net benefit) and performance criteria is evaluated by a system simulation. A system (or systems if more than one have very similar performance) is selected that maximizes the contribution towards the formulation objectives (those that exhibit the highest value of net benefits while satisfying the minimum performance criteria). To confirm the incremental justification of each component, the contribution of each system component in the last added position is evaluated. The last added value is the difference between the value (net benefits) of the system with all components in operation and the value (net benefits) of the system with the last added component removed. If each component is incrementally justified, as indicated by the test, the system is economically justified, and formulation is complete. If any components are not incrementally justified, they should be dropped and the last added analysis repeated.

(2) The system selected by this strategy will be a feasible system that is economically justified. Assuming the method of devising the alternative systems is rational, the chances are good that the major worthwhile projects will have been identified. On the other hand, the chances that this system provides the absolute maximum net benefits is relatively small. This strategy would require between 30 and 60 system evaluations for a moderately complex (15 component) system.

d. First added strategy. This strategy is designed such that its successive application will yield the formulated system. The performance of the systems, including the base components (if any), are evaluated with each potential addition to the system in the "first added" position. The component that contributes the greatest value (net benefit) to the system is selected and added to the base system.

(1) The analysis is then repeated for the next stage by computing the first added value of each component to the system again, the base now including the first component added. The strategy is continued to completion by successive application of the first added analysis until no more component additions to the system are justified.

(2) The strategy does have a great deal of practical appeal and probably would accomplish the important task of identifying the components that are clearly good additions to the system and that should be implemented at an early stage. The strategy, however, ignores any system value that could be generated by the addition of more than one component to the system at a time, and this could omit potentially useful additions to the system. For example, the situation sometimes exists where reservoirs on, say, two tributaries above a damage center are justified, but either one analyzed separately is not, i.e., the system effect is great enough to justify both. The number of system analyses required to formulate a system based on this strategy could range upwards to 120 for a moderately complex (15 component) system, which is probably close to being an unmanageably large number of evaluations.

e. Last added strategy. This strategy, similar to first added strategy, is designed such that successive application yields the formulated system. Beginning with all proposed components to the system, the value of each component in the last added position is computed. The project whose deletion causes the value (net benefit) of the system to increase the most is dropped out. The net benefits would increase if the component is not incrementally justified. The strategy is continued through successive staged applications until the deletion of a component causes the total system value (net benefits) to decrease.

(1) The last added strategy will also yield a system in which all components are incrementally justified and in which the total system will be justified. This strategy would probably identify the obviously desirable projects, as would the others. However, its weakness is that it is slightly possible, though not too likely, that groups of projects that would not be justified are carried along because of their complex linkage with the total system. For example, the situation sometimes exists where reservoirs on two tributaries above a damage center are not justified together, but deletion of each from a system that includes both results in such a great loss in system value that individual analysis indicates neither should be dropped individually.

(2) The number of systems analyses required for this strategy would be similar to the first added strategy

requiring perhaps 10-20 percent more evaluations. Twenty-two last added analyses were made in the four stages required to select four new projects out of seven alternatives. This strategy is more efficient than the first added if the majority of the potential system additions are good ones.

f. Branch-and-bound enumeration. "Branch-and-bound enumeration is a general-purpose technique for identifying the optimal solution to an optimization problem without explicitly enumerating all solutions," (HEC 1985a). The technique provides a framework to evaluate independent alternatives by dividing the entire set into subsets for evaluation. The method has been applied in resource planning to problems of sizing, selecting, sequencing, and scheduling projects. HEC has developed a training document illustrating the application to flood-damage-mitigation plan selection (HEC 1985). Additionally, HEC Research Document No. 35 (Bowen 1987) illustrates an application for reservoir flood control plan selection using computer program HEC-5 for reservoir simulation. The procedure can use any criteria for evaluation and supports detailed simulation in the analysis process.

4-11. General Study Procedure

There is no single approach to developing an optimum plan of improvement for a complex reservoir system. Ordinarily many services are fixed and act as constraints on system operation for other services. In many cases, all but one service is fixed, and the system is planned to optimize the output for one remaining service, such as power generation. It should also be recognized that most systems have been developed over a long period of time and that many services are in fact fixed, as are many system features. Nevertheless, an idealized general study procedure is presented below:

a. Prepare regional and river-system topographic maps showing locations of hydrologic stations, existing and contemplated projects, service and damage areas, and pertinent drainage boundaries. Obtain all precipitation, evaporation, snowpack, hydrograph timing and runoff data pertinent to the project studies. Obtain physical and operational data on existing projects. Construct a normal seasonal isohyetal map for the river basin concerned.

b. For each location where flood protection is to be provided, estimate approximately the nondamaging flow capacity that exists or could be ensured with minor channel and levee improvements. Estimate also the amount of storage (in addition to existing storage) that would be needed to provide a reasonable degree of protection, using procedures described in Chapter 10. Distribute this storage in a reasonable way among contemplated reservoirs in order to obtain a first approximation of a plan for flood control. Include approximate rule curves for releasing some or all of this storage for other uses during the nonflood season where appropriate.

c. Determine approximately for each tributary, where appropriate, the total water needed each month for all conservation purposes and attendant losses, and, using procedures described in Chapter 11, estimate the storage needed on each principle tributary for conservation services. Formulate a basic plan of development including detailed specification of all reservoir, canal, channel, and powerplant features and operation rules; all flow requirements; benefit functions for all conservation services; and stage-damage functions for all flood damage index locations. Although this part of plan formulation is not entirely a hydrologic engineering function, a satisfactory first approximation requires good knowledge of runoff characteristics, hydraulic structure characteristics and limitations, overall hydroelectric power characteristics, engineering feasibility, and costs of various types of structures, and relocations.

d. Using the general procedures outlined in Part 2, develop flood frequencies, hypothetical flood hydrographs, and stage-discharge relations for unregulated conditions and for the preliminary plan of development for flood control. It may be desirable to do this for various seasons of the year in order to evaluate seasonal variation of flood-control space. Evaluate the flood-control adequacy of the plan of development, using procedures described in paragraph 4-5 and Chapter 10, modify the plan, as necessary, to improve the overall net benefits for flood control while preserving basic protection where essential. Each modification must be followed by a new evaluation of net benefits for flood control. Each iteration is costly and time-consuming; consequently, only a few iterations are feasible, and considerable thought must be given to each plan modification.

e. For system analysis to determine the best allocation of flow and storage for conservation purposes, consider optimization using a tool HEC-PRM (paragraph 4-9c). The program outputs would then be analyzed to infer an operation policy that could be defined for simulation and more detailed analysis. The alternative is to repeatedly simulate with critical low-flow periods to develop a policy to meet system goals and then perform a period-of-record simulation to evaluate total system performance.

f. Consider generating synthetic sequences of flow to evaluate the system's performance with different flow sequences (see paragraph 5-5). Future system flows replicate the period-of-record. Also, projected changes in the basin should be factored into the analysis. Typically, future conditions are estimated at several stages into the future. The system analysis should be performed for each stage. While these analyses will take additional time and effort, they will also provide some indication of how responsive the system results are to changing conditions.

PART 2

HYDROLOGIC ANALYSIS

Chapter 5
Hydrologic Engineering Data

5-1. Meteorological Data

The extent of meteorological observations is determined by the data needs and use and the availability of personnel and equipment. Data usually recorded at weather stations include air (sometimes water) temperature, precipitation, wind, and evaporation. As indicated below, more extensive recording of various types of data is often made for special purposes. The primary source of meteorological data for the United States is the National Oceanic and Atmospheric Administration (NOAA) National Climatic Data Center in Asheville, North Carolina. Data are available from NOAA in computer-readable form as well as published reports. NOAA publication "Selective Guide to Climatic Data Sources" is an excellent reference for data availability. Private vendor sources employing compact disc (CD) technology using NOAA records are also available.

a. *Storm meteorology.* The determination of runoff potential, particularly flood potential, in areas where hydrologic data are scarce can be based on a knowledge of storm meteorology. This includes sources of moisture in the paths over which the storm has traveled, as well as a knowledge of the mechanics of storm activity. Derivation of hydrologic quantities associated with various storms must take into consideration the type of storm, its path, potential moisture capacity and stability of the atmosphere, isobar, wind and isotherm patterns, and the nature and intensity of fronts separating air masses. These are usually described adequately in the synoptic charts that are prepared at regular (usually 6-hr) intervals for weather forecasting purposes and associated upper-air soundings. Where such charts are available, it is important that they be retained as a permanent record of meteorological activity for use in supplementing information contained in the regularly prepared hemispheric charts. These latter charts summarize the daily synoptic situation throughout the hemisphere but do not contain all of the data that are of interest or that would have direct bearing on the derivation of design criteria.

(1) NOAA monthly publication "Storm Data" (1959-present) documents the time, location, and the meteorologic characteristics of all reported severe storms or unusual weather phenomena. Synoptic maps are published by NOAA on a weekly basis in "Daily Weather Maps, Weekly Series" and on a monthly basis, "Synoptic Weather Maps,

Daily Series, Northern Hemisphere Sea-Level and 500-Millibar Charts and Data Tabulations."

(2) Storm data including synoptic charts for selected historic storms are included in the "Hydrometeorological Reports" and "Technical Memorandum" prepared by the National Weather Service (NWS) Office of Hydrology in Silver Spring, MD. Other sources of meteorological data include the National Hurricane Center in Coral Gables, FL, and state climatologists as well as U.S. Geological Survey and Corps flood reports.

b. *Precipitation.* Monthly summaries of observed hourly and daily precipitation data are published by NOAA in "Climatical Data" and "Hourly Precipitation Data." Precipitation data are also available from NOAA in computer-readable media. Precipitation data for significant historic storms (1870's-1960's) are tabulated in "Storm Rainfall in the United States, Depth, Area, Duration Data."

(1) There are usually local, or state agencies, collecting precipitation data for their own use. These data could provide additional storm information. However, precipitation measurements at remote unattended locations may not be consistently and accurately recorded, particularly where snow and hail frequently occur. For this reason, records obtained at unattended locations must be interpreted with care. When an observer is regularly on-site, the times of occurrence of snowfall and hail should be noted to make accurate use of the data. The exact location and elevation of the gauge are important considerations in precipitation measurement and evaluation. For uniform use, this is best expressed in terms of latitude and longitude and in meters or feet of elevation above sea level. Of primary importance in processing the data is tabulating precipitation at regular intervals. This should be done daily for non-recording gauges with the time of observation stated. Continuously recording gauges should be tabulated hourly. The original recording charts should be preserved in order to permit study of high-intensity precipitation during short intervals for certain applications.

(2) Procedures to develop standard project and probable maximum precipitation estimates are presented in NWS hydrometeorological reports and technical memorandum. Chapter 7 of this manual provides an overview of hypothetical storms and their application to flood-runoff analysis.

c. *Snowpack.* Where snowmelt contributes significantly to runoff, observations of the snowpack characteristics can be of considerable value in the development of hydrologic design criteria. The observation of water equivalent (weight) of a vertical column sampled from the snowpack at specified locations and observation times is of primary importance. As the observations will ordinarily be used as an index for surrounding regions, the elevation and exposure of the location must be known. The depth of snowpack is of secondary importance, but some observations of the areal extent of the snowpack are often useful.

(1) An important element in processing snowpack observation data is the adjustment of observations at all locations to a common date, such as the first of each month during the snowpack accumulation season. Since these observations are often made by traveling survey teams, they are not made simultaneously. Also, they cannot always be made at a specified time because it is impossible to obtain accurate or representative measurements during snowstorms.

(2) Where continuous recordings of snowpack water equivalent by means of radioactive gauges or snow pillows are available, these can be used on a basis for adjusting manual observations at nearby locations. Otherwise, some judgment or correlation technique based on precipitation measurements is required to adjust the observations data to a uniform date at all locations. It is important to preserve the original records whenever such adjustments are made. However, data that are disseminated for use in design should be the adjusted systematic quantities.

(3) The primary agency for the collection and distribution of snowpack data is the Soil Conservation Service (SCS) (Department of Agriculture, Washington, D.C.). From January through May each year, the SCS publishes a monthly report titled, "Water Supply Outlook." These reports provide snowpack and streamflow forecast data for each state and region. The SCS also issues "Basin Outlook Reports," a monthly regional summary of snow depth and water content. Additionally, NOAA distributes an annual tabulation of snowpack data in their "Snow Cover Surveys." Climatoligcal precipitation data published monthly by NOAA also include information on snowfall and snow on the ground.

d. Temperatures. In most hydrologic applications of temperature data, maximum and minimum temperatures for each day at ground level are very useful. Continuous records of diurnal temperature variations at selected locations can be used to determine the daily temperature pattern fairly accurately at nearby locations where only the maximum and minimum temperatures are known. In applying temperature data to large areas, it must be recognized that temperatures normally decrease with increasing elevation and latitude. It is also important to preserve all of the original temperature records. Summaries of daily

maximum and minimum temperatures should be maintained and, where feasible, published. The NOAA report series on "Climatography of the United States" by city, state, or region also provides information on daily and monthly normal temperatures.

e. Moisture. Atmospheric moisture is a major factor influencing the occurrence of precipitation. This moisture can be measured by atmospheric soundings which record temperature, pressure, relative humidity, and other items. Total moisture in the atmosphere can be integrated and expressed as a depth of water. During storms, the vertical distribution of moisture in the atmosphere ordinarily follows a rather definite pattern. Total moisture can therefore be related to the moisture at the surface, which is a function of the dew point at the surface. Accordingly, a record of daily dew points is of considerable value. Here, again, the elevation, latitude and longitude of the measuring station must be known. NOAA publication "Local Climatological Data" is a primary source for observed dew point, pressure, and temperature data.

f. Winds. Probably the most difficult meteorological element to evaluate is wind speed and direction. Quite commonly, the direction of surface winds reverses diurnally, and wind speeds fluctuate greatly from hour to hour and minute to minute. There is also a radical change of wind speed and direction with altitude. The speed and direction at lower levels is greatly influenced by obstructions such as mountains, and locally by small obstructions such as buildings and trees. Accordingly, it is important that great care is exercised in selecting a location and altitude for wind measurement. For most hydrologic applications, wind measurements at elevations of 5 to 15 m above the ground surface are satisfactory. It is important to preserve all basic records of winds, including data on the location, ground elevation, and the height of the anemometer above the ground. An anemometer is an instrument for measuring and indicating the force or speed of the wind. Where continuous records are available, hourly tabulations of speed and direction are highly desirable. Total wind movement and the prevailing direction for each day are also useful data. Daily wind data for each state are published in NOAA publications "Local Climatological Data" and "Climatological Data."

g. Evaporation. Evaporation data is usually required for reservoir studies, particularly for low-flow analysis. Reservoir evaporation is typically estimated by measuring pan evaporation or computing potential evaporation. There are several methods of estimating potential evaporation, based on meteorological information. Pruitt (1990) reviewed various approaches in an evaluation of the methodology and results published in "A Preliminary

Assessment of Corps of Engineer Reservoirs, Their Purposes and Susceptibility to Drought," (HEC 1990e).

(1) Evaporation is usually measured by using a pan about 4 ft (1.2 m) in diameter filled with water to a depth of about 8 in. (0.2 m). Daily evaporation can be calculated by subtracting the previous day's reading from today's reading and adding the precipitation for the intervening period. The pan should be occasionally refilled and this fact noted in the record. This volume of added water, divided by the area of the pan, is equal to the daily evaporation amount expressed in inches or millimeters. A tabulation of daily evaporation amounts should be maintained and, if possible, published. It is essential that a rain gauge be maintained at the evaporation pan site, and it is usually desirable that temperature, dew point (or wet-bulb temperature) and low-level wind measurements also be made at the site for future study purposes.

(2) NOAA Technical Report "NWS 33, Evaporation Atlas for the Contiguous United States" (Farnsworth, Thompson, and Peck 1982) provides maps showing annual and May-October evaporation in addition to pan coefficients for the contiguous United States. Companion report "NWS 34, Mean Monthly, Seasonal, and Annual Pan Evaporation for the United States" documents monthly evaporation data which was used in the development of the evaporation atlas. Daily observed evaporation data are published for each state in NOAA publications "Local Climatological Data" and in "Climatological Data."

h. Upper air soundings. Upper air soundings are available from NOAA National Climatic Data Center in Asheville, NC. The soundings provide atmospheric pressure, temperature, dew point temperature, wind speed, and direction data from which lapse rate, atmospheric stability, and jet stream strength can be determined. These meteorological parameters are necessary to a comprehensive storm study.

5-2. Topographic Data

a. Mapping. For most hydrologic studies, it is essential that good topographic maps be used. It is important that the maps contain contours of ground-surface elevation, so that drainage basins can be delineated and important features such as slopes, exposure, and stream patterns can be measured. United States Operational Navigation Charts, with a scale of 1:1,000,000 and contour intervals of 1,000 ft, are available for most parts of the world. However, mapping to a much larger scale (1:25,000) and smaller contour intervals in the range of 5-20 ft (1.5 to 6 m) are usually necessary for satisfactory hydrologic studies. The USGS 7.5 Minute Series (Scale 1:24,000), with a typical contour interval of 5 or 10 ft (1.5 or 3.0 m), provides a good basic map for watershed studies. The USGS publications "Catalog of Published Maps" and "Index to Maps" are excellent guides to readily available topographic data for each state. Reports by the USGS are available through Books and Open File Reports Section, USGS, Federal Center, Box 25425, Denver, CO, or by contacting the National Technical Information Service (NTIS), 5285 Port Royal Road, Springfield, VA 22161.

b. Digital mapping. Increasingly, topographic data are available in digital form. One form of computer description of topography is a digital elevation model (DEM). Geographic information systems (GIS) can link land attributes to topographic data and other information concerning processes and properties related to geographic locations. DEM and GIS representations of topologic data are part of a general group of digital terrain models (DTM). Some of the earliest applications in hydrologic modeling used grid cell (raster) storage of information. An example of raster-based GIS is the Corps' Geographic Resource Analysis Support System (GRASS). An alternate approach utilizes a collection of irregularly spaced points connected by lines to produce triangles, known as triangular irregular network (TIN). The use of DEM and TIN data and processing software is rapidly changing and may soon become the standard operation for developing terrain and related hydrologic models. A review of GIS applications in hydrology is provided by DeVantier et al. (HEC 1993).

c. Stream patterns and profiles. Where detailed studies of floodplains are required, computation of water-surface profiles is necessary. Basic data needed for this computation include detailed cross sections of the river and overbank areas at frequent intervals. These are usually obtained by special field surveys and/or aerial photography. When these surveys are made, it is important to document and date the data and resulting models, then permanently preserve the information so it is readily available for future reference. Observations of actual water-surface elevations during maximum flood stages (high-water marks) are invaluable for calibrating and validating models for profile computations.

d. Lakes and swamps. The rate of runoff from any watershed is greatly influenced by the existence of lakes, swamps, and similar storage areas. It is therefore important to indicate these areas on available maps. Data on the outlet characteristics of lakes are important because, in the absence of outflow measurements, the outflow can often be computed using the relationship between the amount of water stored in the lake and its outlet characteristics.

e. Soil and geology. Certain maps of soils and geology can be very useful in surface-water studies if they show characteristics that relate to perviousness of the basin. These can be used for estimating loss rates during storms. Of particular interest are areas of extensive sandy soils that do not contribute to runoff and areas of limestone and volcanic formations that are highly pervious and can store large amounts of water beneath the surface in a short time. Additionally, watershed sediment yield estimates will depend on similar information. The SCS soil survey reports are the primary source of soil and permeability data. State geologic survey or mineral resource agencies are also a useful source of geologic data.

f. Vegetal cover. Often the type of vegetation more accurately reflects variation in hydrologic phenomena than does the type of soil or the geology. In transposing information to areas of little or no hydrologic data, generalized maps of vegetal cover are very helpful. As with soil and geology, vegetation has a significant impact on sediment yield. In the arid southwest, time since the watershed last burned is a significant parameter in estimating total sediment yield for a storm event. The U.S. Forest Service, the Agricultural Stabilization and Conservation Service (ASCS) and, in western states, the Bureau of Land Management are sources of vegetal cover maps. State forest, agricultural agencies, or USGS topographic maps also provide information on vegetal cover.

g. Existing improvements. Streamflow at any particular location can be greatly affected by hydraulic structures located upstream. It is important, therefore, that essential data be obtained on all significant hydraulic structures located in and upstream from a study area. For diversion structures, detailed data are required on the size of the diversion dam, capacity of the diversion canal, and the probable size of flood required to wash out the diversion dam. In the case of storage reservoirs, detailed data on the relation of storage capacity to elevation, location, and size of outlets and spillways, types, sizes, and operation of control gates, and sizes of power plant and penstocks should be known. Bridges can produce backwater effects which will cause upstream flooding. This flooding may be produced by the approach roads, constriction of the channel and floodplain, pier shapes, the angle between the piers and the streamflow, or the pier length-width ratios.

5-3. Streamflow Data

The availability of streamflow data is a significant factor in the selection of an appropriate technical method for reservoir studies. It is important to be cognizant of the nature, source, reliability, and adequacy of available data. If estimates are needed, the assumptions used should be documented, and the effect of errors in the estimates on the technical procedure and results should be considered.

a. Measurement. Streamflow data are usually best obtained by means of a continuous record of river stage, supplemented by frequent meter measurements of flows that can be related to corresponding river stages. It is important that stage measurements be made at a good control section, even if a weir or other control structures must be constructed. Each meter measurement should consist of velocity measurements within each of several (6-20, where practical) subdivisions within the channel cross section. Velocity for a subdivision is usually taken as the velocity at a depth of 60 percent (0.6) of the distance from the surface to the streambed or as the average of velocities taken at 20 and 80 percent (0.2 and 0.8) of the depth at the middle of the subdivision. River stage readings should be made immediately before and after the cross section is metered. The average of these two stages is the stage associated with the measurement. The measurement is computed by integrating the rates of flow (m^3/s) in all subdivisions of the cross section.

(1) Measurements of stream velocity and computed streamflow are usually recorded on standard forms. When measurements have been made for a sufficient range of flows, the rating curve of flow versus stage can be developed. The rating curve can be used to convert the continuous record of stage into a continuous record of flow. The flows should be averaged for each day in order to construct a tabulation of mean daily flows. This constitutes the most commonly published record of runoff.

(2) For flood studies, it is particularly important to obtain accurate records of short-period variations during high river stages and to obtain meter measurements at or near the maximum stage during as many floods as possible. Where the river profile is very flat, as in estuaries and major rivers, it is advisable to obtain measurements frequently on the rising and on the falling stage to determine if a looped, or hysteresis, effect exists in the rating. The reason for this is that the hydraulic slope can change greatly, resulting in different rating curves for rising and falling stages.

b. Streamflow data sources. The USGS is the primary agency for documenting and publishing flow data in the United States. Daily flow data for each state are published in the USGS annual "Water Data Report." The USGS National Water Data Exchange (NAWDEX) computerized database identifies sources of water data. The National Water Data Storage and Retrieval System (WATSTORE) provides processing, storage, and retrieval of surface water, groundwater, and water quality data.

NAWDEX is only an index of the contents of WATSTORE. These programs will eventually be integrated into a National Water Information System, which will also combine the National Water-Use Information Program and Water Resources Scientific Information Center (Mosley and McKerchar 1993). Commercial data services have also provided convenient access to USGS daily and peak flow files on CD.

c. Flow conditions. Reservoirs substantially alter the distribution of flow in time. Many other developments, such as urbanization, diversions, or cultivation and irrigation of large areas can also have a significant effect on watershed yield and the distribution of flow in time. The degree that flows are modified depends on the scale and manner of the development, as well as the magnitude, time, and areal distribution of rainfall (and snowmelt, if pertinent). Most reservoir evaluations require an assessment based on a consistent flow data set. Various terms are used to define what condition the data represent:

(1) Natural conditions in the drainage basin are defined as the hydrologic conditions that would prevail if no regulatory works or other development affecting basin runoff and streamflow were constructed. The effects of natural lakes and swamp areas are included.

(2) Present conditions are defined as the conditions that exist at, or near, the time of study. If there are upstream reservoirs in the basin, the observed flow record would represent "regulated flow." Flow records, preceding current reservoir projects, would be adjusted to reflect those project operations in order to have consistent "present conditions" flow.

(3) Unregulated conditions reflect the present (or recent) basin development, but without the effect of reservoir regulation. Unlike natural conditions which are difficult to determine, only the effect of reservoir operation and major diversions are removed from the historic data.

(4) Without-project conditions are defined as the conditions that would prevail if the project under consideration were not constructed but with all existing and future projects under construction assumed to exist.

(5) With-project conditions are defined as the conditions that will exist after the project is completed and after completion of all projects having an equal or higher priority of construction.

5-4. Adjustment of Streamflow Data

The adjustment of recorded streamflows is often required before the data can be used in water resources development studies. This is because flow information usually is required at locations other than gauging stations and for conditions of upstream development other than those under which flows occurred historically. In correlating flows between locations, it is important to use "natural" flows (unaffected by artificial storage and diversion) in order that correlation procedures will apply logically and efficiently. In generating flows, natural flows should be used because general frequency functions, characteristic of natural flows, are employed in this process.

a. Natural conditions. When feasible, flow data should be converted to natural conditions. The conversion is made by adding historical storage changes (plus net evaporation) and upstream diversions (less return flows) to historical flows at the gauging stations for each time interval in turn. Under some conditions, it may be necessary to account for differences in channel and overbank infiltration losses, distributary flow diversions, travel times, and other factors.

b. Unregulated conditions. It is not always feasible to convert flows to natural conditions. Often, required data are not available. Also, the hydrologic effects and timing of some basin developments are not known to sufficiently define the required adjustments. An alternative is to adjust the data to a uniform basin condition, usually near current time. The primary adjustments should remove special influences, such as major reservoirs and diversions, that would cause unnatural variations of flow.

c. Reservoir holdouts. The primary effect of reservoir operation is the storage of excess river flow during high-flow periods, and the release of stored water during low-flow periods. The flow adjustment process requires the addition of the change in water stored (hold-outs) in each time step to the observed regulated flow. Holdouts, both positive and negative, are routed down the channel to each gauge and algebraically added to the observed flow. Hydrologic routing methods, typically used for these adjustments, are described in Chapter 9 of EM 1110-2-1417. The HEC Data Storage System (HEC-DSS) software (HEC 1995) provides a convenient data management system and utilities to route flows and add, subtract, or adjust long time-series flow data.

d. Reservoir losses. The nonproject inflow represents the flow at the project site without the reservoir and includes runoff from the entire effective drainage area above the dam, including the reservoir area. Under nonproject conditions, runoff from the area to be inundated by the reservoir is ordinarily only a fraction of the total precipitation which falls on that area. However, under project conditions infiltration losses over the reservoir area are usually minimal during a rainfall event. Thus, practically all the precipitation falling on the reservoir area will appear as runoff. Therefore, the inflow will be greater under project conditions than under nonproject conditions, if inflow is defined as total contribution to the reservoir before evaporation losses are considered. In order to determine the amount of water available for use at the reservoir, evaporation must be subtracted from project inflow. In operation studies, nonproject inflow is ordinarily converted to available water in one operation without computing project inflow as defined above. This is done in one of two ways: by means of a constant annual loss each year with seasonal variation or with a different loss each period, expressed as a function of observed precipitation and evaporation. These two methods are described in the following paragraphs.

(1) Constant annual loss procedure consists of estimating the evapotranspiration and infiltration losses over the reservoir area for conditions without the project, and the evaporation and infiltration losses over the reservoir area with the project. Nonproject losses are usually estimated as the difference between average annual precipitation and average annual runoff at the location, distributed seasonally in accordance with precipitation and temperature variations. These are expressed in millimeters of depth. Under project conditions, infiltration losses are usually ignored, and losses are considered to consist of only direct evaporation from the lake area, expressed in millimeters for each period. The difference between these losses is the net loss due to the project. Figure 5-1 illustrates the differences between nonproject and project losses.

(2) The variable loss approach uses historical records of long-term average monthly precipitation and evaporation data to account for the change in losses due to a reservoir project. This is accomplished by estimating the average runoff coefficient, the ratio of runoff to rainfall, for the reservoir area under preproject conditions and subtracting this from the runoff coefficient for the reservoir area under project conditions. The runoff coefficient for project conditions is usually 1.0, but a lower coefficient may be used if substantial infiltration or leakage from the reservoir is anticipated. The difference between preproject and project runoff coefficients is the net gain expressed as a

ratio of precipitation falling on the reservoir. This is often estimated to be 0.7 for semi-arid regions. This increase in runoff is subtracted from gross reservoir evaporation, often estimated as 0.7 of pan evaporation, to obtain a net loss.

e. Other losses. In final project studies it is often necessary to consider other types of project losses which may be of minor importance in preliminary studies. Often, these losses cannot be estimated until a project design has been adopted. The importance of these losses is dependent upon their relative magnitude. That is, losses of 5 m^3/s might be considered unimportant for a stream which has a minimum average annual flow of 1,500 m^3/s. Such losses, though, would be significant from a stream with a minimum average annual flow of 25 m^3/s. Various types of losses are discussed in the following paragraphs.

(1) The term "losses" may not actually denote a physical loss of water from the system as a whole. Usually, water unavailable for a specific project purpose is called a "loss" for that purpose although it may be used at some other point or for some other purpose. For example, water which leaves the reservoir through a pipeline for municipal water supply or fish hatchery requirements might be called a loss to power. Likewise, leakages through turbines, dams, conduits, and spillway gates are considered losses to hydropower generation, but they are ordinarily not losses to flow requirements at a downstream station. Furthermore, such losses that become available for use below the dam should be added to inflow at points downstream from the project.

(2) Leakage at a dam or in a reservoir area is considered a loss for purposes which are dependent upon availability of water at the dam or in the reservoir itself. These purposes include power generation, pipelines from the reservoir, and any purpose which utilizes pump intakes which are located at or above the dam. As a rule, leakage through, around, or under a dam is relatively small and is difficult to quantify before a project is actually constructed. In some cases, the measured leakage at a similar dam or geologic area may be used as a basis for estimating losses at a proposed project. The amount of leakage is a function of the type and size of dam, the geologic conditions, and the hydrostatic pressure against the dam.

(3) Leakage from conduits and spillway gates is a function of gate perimeter, type of seal, and head on the gate, and it varies with the square root of the head. The amount of leakage may again be measured at existing projects with various types of seals, and a leakage rate computed per meter of perimeter for a given head. This rate may then be used to compute estimated leakage for a

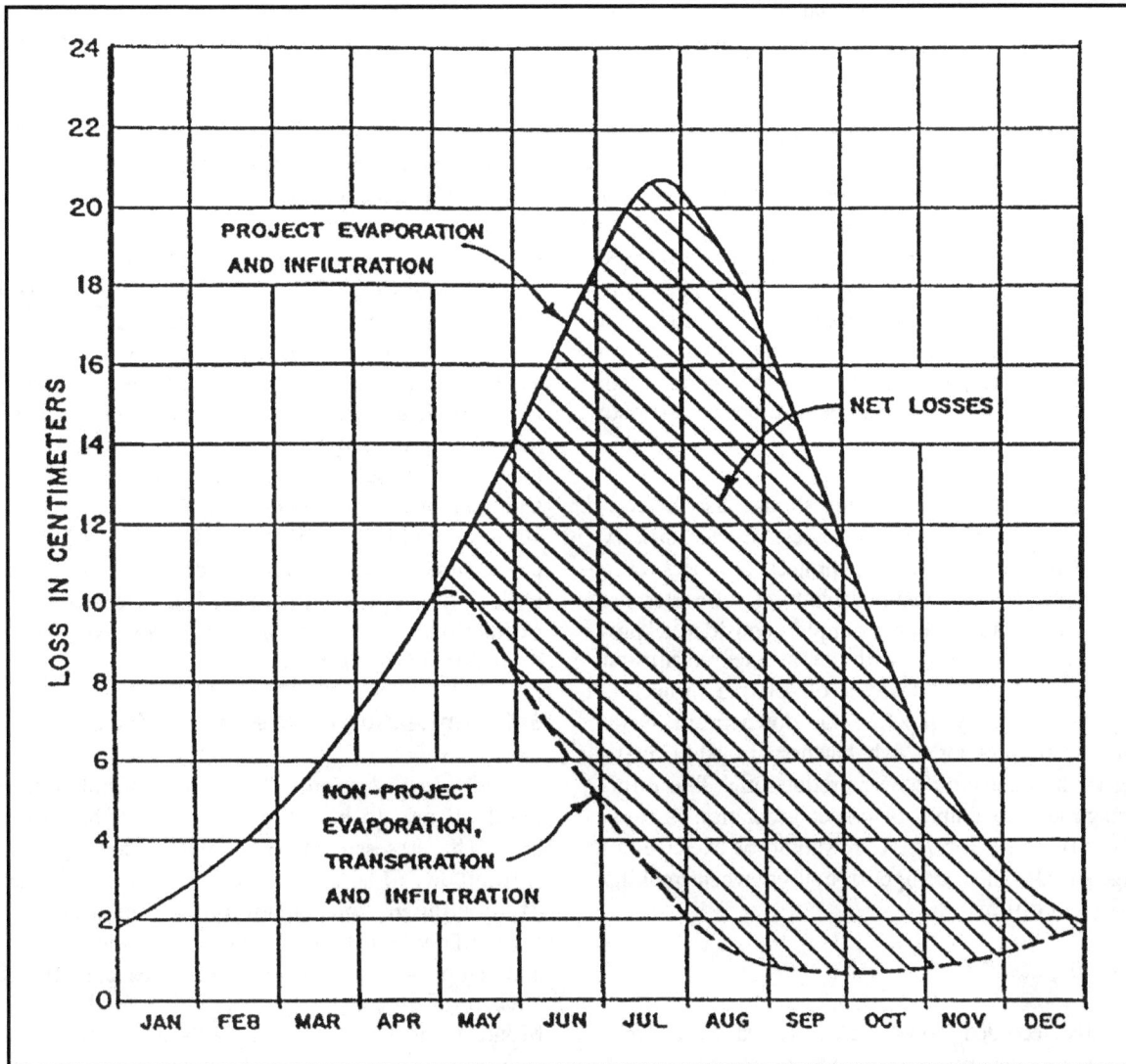

Figure 5-1. Project and nonproject reservoir losses

proposed project by using the proposed size and number of gates and the proposed head on the gates.

(4) If a proposed project will include power, and if the area demand is such that the turbines will sometimes be idle, it is advisable to estimate leakage through the turbines when closed. This leakage is a function of the type of penstock gate, type of turbine wicket gate, number of turbines, and head on the turbine. The actual leakage through a turbine may be measured at the time of acceptance and during annual maintenance inspections, or the measurements of similar existing projects may be used to estimate leakage for a proposed project. An estimate of the percent of time that a unit will be closed may be obtained from actual operational records for existing units in the same demand area. The measured or estimated leakage

rate is then reduced by multiplying by the proportion of time the unit will be closed. For example, suppose that leakage through a turbine has been measured at 0.1 m³/s, and the operation records indicate that the unit is closed 60 percent of the time. The average leakage rate would be estimated at 0.1 × 0.6 or 0.06 m³/s.

(5) The inclusion of a navigation lock at a dam requires that locking operations and leakages through the lock be considered. The leakage is dependent upon the lift or head, the type and size of lock, and the type of gates and seals. Again, estimates can be made from observed leakage at similar structures. Water required for locking operations should also be deducted from water available at the dam site. These demands can be computed by multiplying the volume of water required for a single locking operation

times the number of operations anticipated in a given time period and converting the product to a flow rate over the given period.

(6) Water use for purposes related to project operations is often treated as a loss. Station use for sanitary and drinking purposes, cooling water for generators, and water for condensing operations have been estimated to be about 0.06 m³/s per turbine at some stations in the southwestern United States. Examining operation records for comparable projects in a given study area may also be useful in estimating these losses. If house units are included in a project to supply the project's power requirements, data should be obtained from the designer in order to estimate water used by the units.

(7) The competitive use of water should also be considered when evaluating reservoir losses. When initially estimating yield rates for various project purposes at a multiple-purpose project, competitive uses of water are often treated as losses. For example, consider a proposed project on a stream with an average minimum usable flow of 16 m³/s. The reservoir of this project is to supply 1.5 m³/s by pipeline for downstream water supply and 2.0 m³/s for a fish hatchery in addition to providing for hydroelectric power production. The minimum average flow available for power generation is thus, 16 - (1.5 + 2.0) = 12.5 m³/s. Care should be exercised in accounting for all such competitive uses when making preliminary yield estimates.

f. Missing data.

(1) After recorded flows are converted to uniform conditions, flows for missing periods of record at each pertinent location should be estimated by correlation with recorded flows at other locations in the region. Usually, only one other location is used, and linear correlation of flow logarithms is used. It is more satisfactory, however, to use all other locations in the region that can contribute independent information on the missing data. Although this would require a large amount of computation, the computer program HEC-4 *Monthly Streamflow Simulation* accomplishes this for monthly streamflow (HEC 1971).

(2) Flow estimates for ungauged locations can be estimated satisfactorily on a flow per basin area basis in some cases, particularly where a gauge exists on the same stream. In most cases, however, it is necessary to correlate mean flow logarithms (and sometimes standard deviation of flow logarithms) with logarithms of drainage area size, logarithm of normal seasonal precipitation, and other basin characteristics. Correlation procedures and suggested basin characteristics are described in Chapter 9 of EM 1110-2-1415.

g. Preproject conditions. After project flows for a specified condition of upstream development are obtained for all pertinent locations and periods, they must be converted to preproject (nonproject) conditions. Nonproject conditions are those that would prevail during the lifetime of the proposed project if the project was not constructed. This conversion is made by subtracting projected upstream diversions and storage changes and by accounting for evaporation, return flows, differences in channel infiltration, and timing. Where nonproject conditions will vary during the project lifetime, it is necessary to convert to two or more sets of conditions, such as those at the start and end of the proposed project life. Separate operation studies would then be made for each condition. This conversion to future conditions can be made simultaneously with project operation studies, but a separate evaluation of nonproject flows is usually required for economic evaluation of the project.

5-5. Simulation of Streamflow Data

a. Introduction. The term "simulation" has been used to refer to both the estimation of historic sequences and the assessment of probable future sequences of streamflow. The former reference concerns the application of continuous precipitation-runoff models to simulate streamflow based on meteorologic input such as rainfall and temperature. The latter reference concerns the application of stochastic (probabilistic) models that employ Monte Carlo simulation methods to estimate the probable occurrence of future streamflow sequences. Assessment of the probable reliability of water resource systems can be made given the assessment of probable future sequences of flow. Statistical methods used in stochastic models can also be employed to augment observed historic data by filling in or extending observed streamflow records.

b. Historic sequences from continuous precipitation-runoff models. Many different types of continuous simulation runoff models have been used to estimate the historic sequence of streamflow that would occur from observed precipitation and other meteorologic variables. Among the most prominent are the various forms of the Stanford Watershed Model (Mays and Tung 1992) and the SSARR Model used by the North Pacific Division (USACE 1991). For a further description of the application of the models see EM 1110-2-1417 Section 8.

c. Stochastic streamflow models. Stochastic stream-flow models are used to assess the probable sequence of future flows. As with any model, a model structure is assumed, parameters are estimated from observed data, and the model is used for prediction (Salas et al. 1980). Typically, stochastic streamflow models are used to simulate annual and/or monthly streamflow volumes. Stochastic streamflow models have not been successfully developed for daily streamflow.

(1) Although many different structures have been proposed in the research literature, regression is most commonly used as the basis for stochastic streamflow models. The regressions involve both correlation between flows at different sites and the correlation between current and past flow periods, termed serial correlation. The correlation between sites is useful in improving parameters estimates from regional information. The serial correlation between periods is important in modeling the persistence, or the tendency for high flow or drought periods. A random error component is added to the regression to provide a probabilistic component to the model.

(2) The model parameters are estimated to preserve the correlation structure observed in the observed data. If the appropriate correlation structure is preserved, then the regression residuals should closely approximate the behavior that would be expected from a random error component.

(3) Model prediction is performed via the application of Monte Carlo simulation. Monte Carlo simulation is a numerical integration technique. This numerical technique is necessary because the stochastic model effectively represents a complex joint probability distribution of streamflows in time and space that cannot be evaluated analytically. The simulation is performed by producing random sequences of flows via a computer algorithm that employs random number generators. These sequences of flows are analyzed to assess supply characteristics, for example the probability for a certain magnitude or duration of drought. The number of flow sequences generated is sufficient when the estimated probabilities do not change significantly with the number of simulations. For further explanation of this point, see "Stochastic Analysis of Drought Phenomena" (HEC 1985b).

d. Assessment of reliability with stochastic stream-flow models. The advantage of using a stochastic streamflow model over that of employing only historic records is that it can be used to provide a probabilistic estimate of a water resource system's reliability. For example, the probability that a particular reservoir will be able to meet certain goals can be estimated by simulating the stochastic flow sequences with a reservoir simulation model. Once again, the number of flow sequences used are sufficient when the estimate of the probabilities stabilize.

e. Available software for stochastic streamflow simulation. HEC-4, "Monthly Streamflow Simulation" (HEC 1971), and LAST (Lane 1990) are public domain software for performing stochastic streamflow simulation. HEC-4 performs monthly stochastic streamflow simulation. LAST utilizes a more modern approach where annual and shorter time period (seasonal, monthly, etc.) stochastic streamflow can be co-simulated.

f. Extending and filling in historic records. Statistical techniques can be used to augment existing historic records by either "filling in" missing flow values or extending the observed record at a gauge based on observations at other gauges. The statistical techniques used are referred to as MOVE, maintenance of variance extension, and are a modification of regression based techniques (Alley and Burns 1983, and Salas 1992). MOVE algorithms have been instituted because the variance of series augmented by regression alone is underestimated. The MOVE technology is only generally applicable when serial correlation does not exist in the streamflow records. However, monthly or annual sequences of streamflow volumes usually do exhibit a degree of serial correlation. In these circumstances, the information provided by the longer record station may not be useful in extending a shorter record station. For a discussion of the impact of serial correlation see Matalas and Langbein (1962) and Tasker (1983).

Chapter 6
Hydrologic Frequency Determinations

6-1. Introduction

Frequency curves are most commonly used in Corps of Engineers studies to determine the economic value of flood reduction projects. Reservoir applications also include the determination of reservoir stage for real estate acquisition and reservoir-use purposes, the selection of rainfall magnitude for synthetic floods, and the selection of runoff magnitude for sizing flood-control storage.

a. Annual and partial-duration frequency. There are two basic types of frequency curves used in hydrologic work. A curve of annual maximum events is ordinarily used when the primary interest lies in the very large events or when the second largest event in any year is of minor concern in the analysis. The partial-duration curve represents the frequency of all events above a given base value, regardless of whether two or more occurred in the same year. This type of curve is ordinarily used in economic analysis when there are substantial damages resulting from the second largest and third largest floods in extremely wet years. Damage from floods occurring more frequently than the annual event can occur in agricultural areas, when there is sufficient time between events for recovery and new investment. When both the frequency curve of annual floods and the partial-duration curve are used, care must be exercised to assure that the two are consistent.

b. Seasonal frequency curves. In most locations, there are seasons when storms or floods do not occur or are not severe, and other seasons when they are more severe. Also, damage associated with a flood often varies with the season of the year. In studies where the seasonal variation is of primary importance, it becomes necessary to establish frequency curves for each month or other subdivision of the year. For example, one frequency curve might represent the largest floods that occur each January; a second one would represent the largest floods that occur each February, etc. In another case, one frequency curve might represent floods during the snowmelt season, while a second might represent floods during the rainy season. Occasionally, when seasons are studied separately, an annual-event curve covering all seasons is also prepared. Care should be exercised to assure that the various seasonal curves are consistent with the annual curve.

6-2. Duration Curves

a. Flow duration curve. In power studies, for run-of-river plants particularly and in low-flow studies, the flow-duration curve serves a useful purpose. It simply represents the percent of time during which specified flow rates are exceeded at a given location. Ordinarily, variations within periods less than 1 day are inconsequential, and the curves are therefore based on observed mean-daily flows. For the purposes served by flow duration curves, the extreme rates of flow are not important, and consequently there is no need for refining the curve in regions of high flow.

b. Preparing flow-duration curve. The procedure ordinarily used to prepare a flow duration curve consists of counting the number of mean-daily flows that occur within given ranges of magnitude. The lower limit of magnitude in each range is then plotted against the percentage of days of record that mean-daily flows exceed that magnitude. A flow duration curve example is shown in Figure 6-1.

6-3. Flood-Frequency Determinations

At many locations, flood stages are a unique function of flood discharges for most practical purposes. Accordingly, it is usual practice to establish a frequency curve of river discharges as the basic hydrologic determination for flood damage reduction project studies. In special cases, factors other than river discharge, such as tidal action or accumulated run-off volume, may greatly influence river stages. In such cases, a direct study of stage frequency based on recorded stages is often warranted.

a. Determination made with available data. Where runoff data at or near the site are available, flood-frequency determinations are most reliably made by direct study of these data. Before frequency studies of recorded flows are made, the flows must be converted to a uniform condition, usually to conditions without major regulation or diversion. Developing unregulated flow requires detailed routing studies to remove the effect of reservoir hold-outs and diversions. As damaging flows occur during a very small fraction of the total time, only a small percentage of daily flows are used for flood-frequency studies. These consist of the largest flow that occurs each year and the secondary peak flows that cause damage. However, for most reservoirs studies, the period-of-record flow will be required for analysis of nonflood purposes and impacts.

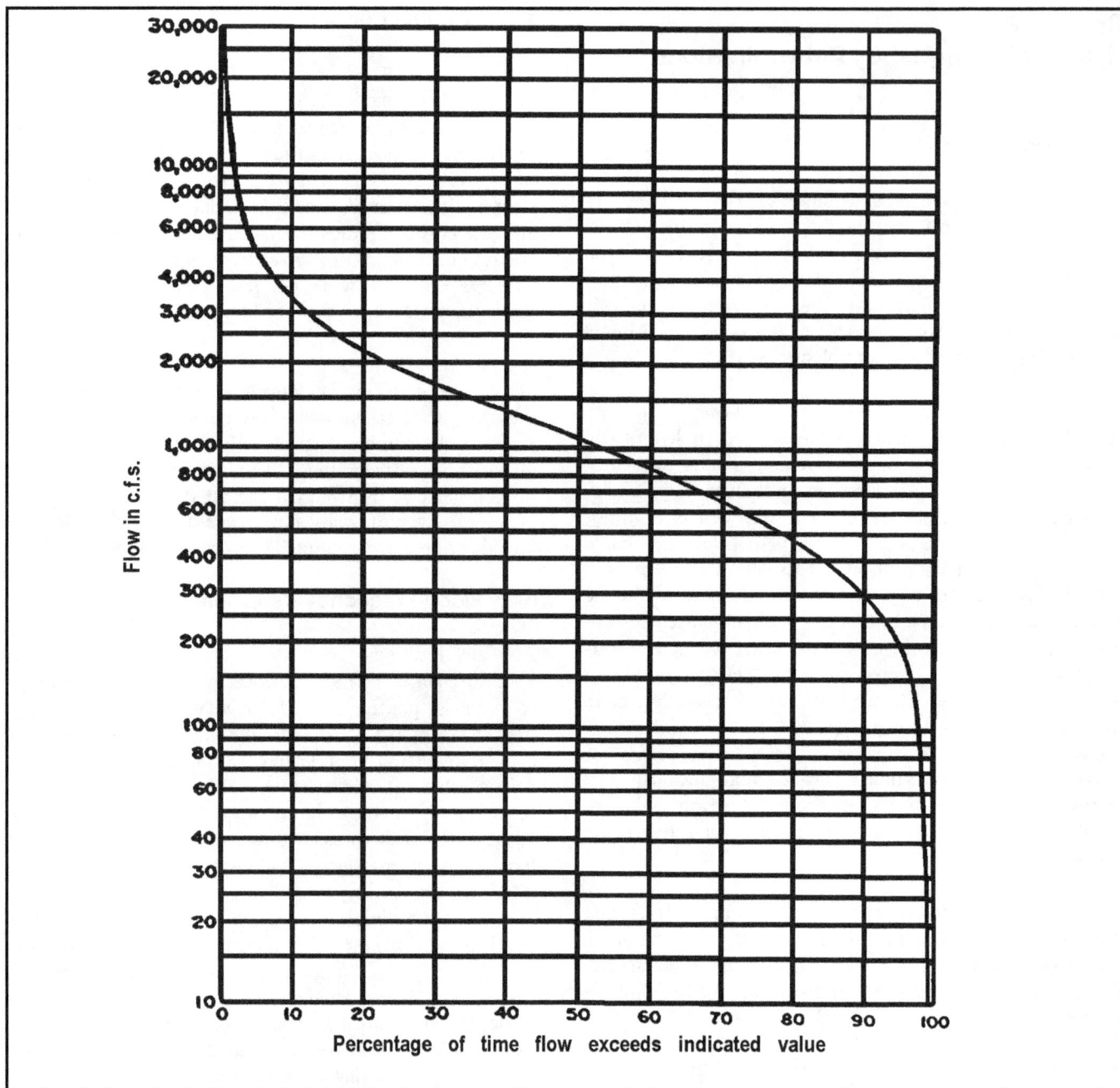

Figure 6-1. Example flow duration curve

b. Historical data. Flood frequency estimates are subject to considerable uncertainty, even when fairly long records are available. In order to increase the reliability of frequency estimates, empirical theoretical frequency relations are used in specific frequency studies. These studies require that a complete set of data be used. In order to comply with this requirement, the basic frequency study ordinarily is based on the maximum flow for each year of record. Supplementary studies that include other damaging events are ordinarily made separately. The addition of historical information can be very important in verifying the frequency of large recorded events. Historical information on large damaging floods can be obtained through standard sources such as USGS water supply series or from newspapers and local museums. The latter sources often are more qualitative but give important insight into the relative frequency of recorded events.

c. Selecting and computing frequency curves. The underlying general assumption made in all frequency studies is that each observed event represents an approximately equal proportion of the future events that will occur at the location, if controlling conditions do not change. Detailed procedures for selecting data and computing flood-frequency curves are presented in EM 1110-2-1415 and HEC-FFA program (HEC 1992c).

d. Regional correlation of data. Where runoff data at or near the site do not exist or are too fragmentary to support direct frequency calculations, regional correlation of frequency statistics may be used for estimating frequencies. These correlations generally relate the mean and standard deviation of flows to drainage basin characteristics and location. Techniques of regional correlation are presented in Chapter 9 of EM 1110-2-1415.

e. Extreme floods. In the analysis of reservoir projects, the project's performance during floods larger than the maximum recorded events is usually required. Extrapolating derived frequency relations is uncertain, so special studies of the potential magnitudes of extreme flood events are usually required. The most practical approach is through examination of rainstorms that have occurred in the region and determination of the runoff that would result at the project location if these storms should occur in the tributary area. This subject is discussed in the following chapter, "Flood-Runoff Analysis."

6-4. Estimating Frequency Curves

a. Approaches. There are two basic approaches to estimating frequency curves--graphical and analytical. Each approach has several variations, but the discussion herein will be limited to recommended methods. The primary Corps reference for computing frequency curves is EM 1110-2-1415.

(1) Graphical. Frequencies are evaluated graphically by arranging observed values in the order of magnitude and representing frequencies by a smooth curve through the array of values. Each observed value represents a fraction of the future possibilities and, when plotting the frequency curve, it is given a plotting position that is calculated to give it the proper weight. Every derived frequency relation should be plotted graphically, even though the results can be obtained analytically. Paragraph 2-4 of EM 1110-2-1415 presents "Graphical Frequency Analysis."

(2) Analytical. In the application of analytical (statistical) procedures, the concept of theoretical populations or distributions is employed. The events that have occurred are presumed to constitute a random sample and are used

accordingly to make inferences regarding their "parent population" (i.e., the distribution from which they were derived). The procedure is applied to annual maxima of unregulated flow, which are assumed to be independent random events. The fact that the set of observations could result from any of many sets of physical conditions or distributions leaves considerable uncertainty in the derived curve. However, using statistical processes, the most probable nature of the distribution from which the data were derived can be estimated. Because this in all probability is not the true parent population, the relative chance that variations from this distribution might be true must be evaluated. Each range of possible parent populations is then weighted in proportion to its likelihood in order to obtain a weighted average. A probability obtained from this weighted average is herein referred to as the expected probability P_N. Chapter 3 of EM 1110-2-1415 covers analytical flood-frequency analysis.

(3) Regional. Because of the shortness of hydrologic records, frequency determinations for rare events are relatively unreliable when based on a single record. Also, it is often necessary to estimate frequencies for locations where no record exists. For these reasons, regionalized frequency studies, in which frequency characteristics are related to drainage-basin features and precipitation characteristics, are desirable. Regionalized frequency studies usually develop relationships for analytical frequency statistics. An alternative approach is to develop predictive equations for the flow for specific recurrence intervals. Chapter 9 of EM 1110-2-1415 presents regression analysis and its application to regional studies.

b. Flood volume-duration frequencies. A comprehensive flood volume-duration frequency series consists of a set of: 1, 3, 7, 15, 30, 60, 120, and 183-day average flows for each water year. These durations are normally available from the USGS WATSTORE files, and they are the default durations in the computer program STATS (HEC 1987a). Runoff volumes are expressed as average flows in order that peak flows and volumes can be readily compared and coordinated. Paragraph 3-8 of EM 1110-2-1415 covers flood volumes.

c. Low flow frequencies. Reservoir analysis often requires the evaluation of the frequencies of low flows for various durations. The same fundamental procedures can be used, except that minimum instead of maximum runoff values are selected from the basic data. For low flows, the effects of basin development are usually more significant than for high flow. A relatively moderate diversion may not be very significant during a flood; however, it may greatly modify or even eliminate low flows. Accordingly, one of the most important aspects of low flows concerns the evaluation of past and future effects of basin developments. Chapter 4 of EM 1110-2-1415 describes low-flow frequency analysis.

d. Reservoir level frequency. A reservoir frequency curve of annual maximum storage is ordinarily constructed graphically, using the procedures for flood-flow frequency. Observed storage should be used to the extent available, but only if the reservoir has been operated in the past in accordance with future plans. If historical data are not available, or if it is not appropriate for future use, then reservoir routings should be used to develop data for expected reservoir operations. Stage-duration curves can be constructed from historical operation data or from simulations. These curves are usually constructed for the entire period-of-record, or for a selected wet or dry period. For some purposes, particularly recreational use, the seasonal variation of reservoir stages is of importance, and a set of frequency or duration curves for each month of the year may be required. Reservoir stage (or elevation) curves should indicate significant reservoir levels such as: minimum pool, top of conservation pool, top of flood-control pool, spillway crest elevation, and top of dam.

6-5. Effect of Basin Developments on Frequency Relations

a. Effects of flood-control works. Most hydrologic frequency estimates serve some purpose relating to the planning, design, or operation of water resources management projects. The anticipated effects of a project on flooding can be assessed by comparing the peak discharge and volume frequency curves with and without the project. Also, projects that have existed in the past have affected the rates and volumes of floods, and recorded values must be adjusted to reflect uniform conditions in order for the frequency analysis to conform to the basic assumptions of randomness and common population. For a frequency curve to conform reasonably with a generalized mathematical or probability law, the flows must be essentially unregulated by man-made storage or diversion structures. Consequently, wherever practicable, recorded runoff values should be adjusted to unregulated conditions before a frequency analysis is made. However, in cases where the regulation results from a multitude of relatively small hydraulic structures that have not changed appreciably during the period of record, it is likely that the general mathematical laws will apply as in the case of natural flows, and that adjustment to natural conditions would be unnecessary. The effects of flood control works are presented in paragraph 3-9 of EM 1110-2-1415, and effects of urbanization in paragraph 3-10.

b. Regulated runoff frequency curves. If it is practical, the most complete approach to determining frequency curves of regulated runoff consists of routing flows for the entire period of record through the proposed management works, arranging the annual peak regulated flows in order of magnitude, assigning a plotting position to the peak values, plotting the peak flow values at the assigned plotting position, and drawing the frequency curve based on the plotted data. A less involved method consists of routing the largest floods of record, or multiples of a large hypothetical flood, to estimate the regulated frequency curve. This approach requires the assumption that the frequency of the regulated peak flow is the same as the unregulated peak flow, which is probably true for the largest floods. Paragraph 3-9*d* of EM 1110-2-1415 describes these methods.

c. Erratic stage-discharge frequency curves. In general, cumulative frequency curves of river stages are determined from frequency curves of flow. In cases where the stage-discharge relation is erratic, a frequency curve of stages can be derived directly from stage data. Chapter 6 of EM 1110-2-1415 presents stage-frequency analysis.

d. Reservoir or channel modifications. Project construction or natural changes in streambed elevation may change the relationship between stage and flow at a location. By forming constrictions, levees may raise river stages half a meter for some distance upstream. Reservoir or channel modifications may cause changes in degradation or aggradation of streambeds, and thereby change rating curves. Thus, the effect of projects on river stages often involves the effects on channel hydraulics as well as the effects on streamflow. Consult EM 1110-2-1416 for information on modeling these potential changes.

6-6. Selection of Frequency Data

a. Primary considerations. The primary consideration in selecting an array of data for a frequency study is the objective of the frequency analysis. If the frequency curve that is developed is to be used for estimating damages that are related to instantaneous peak flows in a stream, peak flows should be selected from the record. If the damages are related to maximum mean-daily flows or to maximum 3-day flows, these items should be selected. If the behavior of a reservoir under investigation is related to the 3-day or 10-day rain-flood volume, or to the seasonal snowmelt volume, that pertinent item should be selected. Normally, several durations are analyzed along with peak flows to develop a consistent relationship.

b. Selecting a related variable. Occasionally, it is necessary to select a related variable in lieu of the one desired. For example, where mean daily flow records are more complete than the records of peak flows, it may be desirable to derive a frequency curve of mean-daily flows and then, from the computed curve, derive a peak-flow curve by means of an empirical relation between mean daily flows and peak flows. All reasonably independent values should be selected, but the annual maximum events should ordinarily be segregated when the application of analytical procedures is contemplated.

c. Data selected. Data selected for a frequency study must measure the same aspect of each event (such as peak flow, mean-daily flow, or flood volume for a specified duration), and each event must be controlled by a uniform set of hydrologic and operational factors. For example, it would be improper to combine items from old records that are reported as peak flows but are in fact only daily readings, with newer records where the peak was actually measured. Similarly, care should be exercised when there has been significant change in upstream storage regulation during the period of record so as not to inadvertently combine unlike events into a single series. In such a case, the entire flow record should be adjusted to a consistent condition, preferably the unregulated flow condition.

d. Hydrologic factors. Hydrologic factors and relationships operating during a winter rain flood are usually quite different from those operating during a spring snowmelt flood or during a local summer cloudburst flood. Where two or more types of floods are distinct and do not occur predominantly in mutual combinations, they should not be combined into a single series for frequency analysis. They should be considered as events from different parent populations. It is usually more reliable in such cases to segregate the data in accordance with type and to combine only the final curves, if necessary. For example, in the mountainous region of eastern California, frequency studies are made separately for rain floods, which occur principally from November through March, and for snowmelt floods, which occur from April through July. Flows for each of these two seasons are segregated strictly by cause, those predominantly caused by snowmelt and those predominantly caused by rain. In desert regions, summer thunderstorms should be excluded from frequency studies of winter rain flood or spring snowmelt floods and should be considered separately. Similarly, in coastal regions it would be desirable to separate floods induced by hurricanes or typhoons from other general flood events.

e. Data adjustments. When practicable, all runoff data should be adjusted to unregulated hydrologic conditions before making the frequency study because natural flows are better adapted to analytical methods and are more easily compared within a region. Frequency curves of present-regulated conditions (those prevailing under current practices of regulation and diversion) or of future-regulated conditions can be constructed from the frequency curve of natural flow by means of an empirical or logical relationship between natural and regulated flows. Where data recorded at two different locations are to be combined for construction of a single frequency curve, the data should be adjusted as necessary to a single location, usually the location of the longer record, accounting for differences of drainage area and precipitation and, where appropriate, channel characteristics between the locations. Where the stream-gauge location is somewhat different from the project location, the frequency curve should be constructed for the stream-gauge location and subsequently adjusted to the project location.

f. Runoff record interruptions. Occasionally, a runoff record may be interrupted by a period of one or more years. If the interruption is caused by the destruction of the gauging station by a large flood, failure to fill in the record for that flood would have a biasing effect, which should be avoided. However, if the cause of the interruption is known to be independent of flow magnitude, the entire period of interruption should be eliminated from the frequency array, since no bias would result. In cases where no runoff records are available on the stream concerned, it is possible to estimate the frequency curve as a whole using regional generalizations. An alternative method is to estimate a complete series of individual floods from recorded precipitation by continuous hydrologic simulation and perform conventional frequency analysis on the simulated record.

6-7. Climatic Variations

Some hydrologic records suggest regular cyclic variations in precipitation and runoff potential. Many attempts have been made to demonstrate that precipitation or stream flows display variations that are in phase with various cycles, particularly the well-established 11-year sunspot cycle. There is no doubt that long duration cycles or irregular climatic changes are associated with general changes of land masses and seas and with local changes in lakes and swamps. Also, large areas that have been known to be fertile in the past are now arid deserts, and large temperate regions have been covered with glaciers one or more times. Although the existence of climatic changes is not questioned, their effect is ordinarily neglected because long-term climatic changes generally have insignificant effects during the period concerned in water development projects, and short-term climatic changes tend to be self-compensating. For these reasons, and because of the difficulty in differentiating between fortuitous and systematic changes, it is considered that, except for the annual cycle, the effect of natural cycles or trends during the period of useful project life can ordinarily be neglected in hydrologic frequency studies.

6-8. Frequency Reliability Analyses

a. Influences. The reliability of frequency estimates is influenced by the amount of information available, the variability of the events, and the accuracy with which the data were measured.

(1) In general with regard to the amount of information available, errors of estimate are inversely proportional to the square root of the number of independent items contained in the frequency array. Therefore, errors of estimates based on 40 years of record would normally be half as large as errors of estimates based on 10 years of record, other conditions being the same.

(2) The variability of events in a record is usually the most important factor affecting the reliability of frequency estimates. For example, the ratio of the largest to the smallest annual flood of record on the Mississippi River at Red River Landing, LA, is about 2.7; whereas the ratio of the largest to the smallest annual flood of record on the Kings River at Piedra, CA, is about 100 or 35 times as great. Statistical studies show that as a consequence of this difference in variability, a flow corresponding to a given frequency that can be estimated within 10 percent on the Mississippi River, can be estimated only within 40 percent on the Kings River.

(3) The accuracy of data measurement normally has relatively little influence on the reliability of a frequency estimate, because such errors ordinarily are not systematic and tend to cancel. The influence of extreme events on reliability of frequency estimates is greater than that of measurement errors. For this reason, it is usually better to include an estimated magnitude for a major flood than to ignore it. For example, a flood event that was not recorded because of gauge failure should be estimated, rather than to omit it from the frequency array. However, it is advisable to always use the most reliable sources of data and to guard against systematic errors.

b. Errors in estimating flood frequencies. It should be remembered that possible errors in estimating flood frequencies are very large, principally because of the chance of having a nonrepresentative sample. Sometimes the occurrence of one or two rare flood events can change the apparent exceedance frequency of a given magnitude from once in 1,000 years to once in 200 years. Nevertheless, the frequency-curve technique is considerably better than any other tool available for certain purposes and represents a substantial improvement over using an array restricted to observed flows only. Reliability criteria useful for illustrating the accuracy of frequency determinations are described in Chapter 8 of EM 1110-2-1415.

6-9. Presentation of Frequency Analysis Results

Information provided with frequency curves should clearly indicate the scope of the studies and include a brief description of the procedure used, including appropriate references. When rough estimates are adequate or necessary, the frequency data should be properly qualified in order to avoid misleading conclusions that might seriously affect the project plan. A summary of the basic data consisting of a chronological tabulation of values used and indicating sources of data and adjustments made would be helpful. The frequency data can also advantageously be presented in graphical form, ordinarily on probability paper, along with the adopted frequency curves.

Chapter 7
Flood-Runoff Analysis

7-1. Introduction

Flood-runoff analysis is usually required for any reservoir project. Even without flood control as a purpose, a reservoir must be designed to safely pass flood flows. Rarely are there sufficient flow records at a reservoir site to meet all analysis requirements for the evaluation of a reservoir project. This chapter describes the methods used to analyze the flood hydrographs and the application of hypothetical floods in reservoir projects. Most of the details on methods are presented in EM 1110-2-1417. The dam safety standards are dependent on the type and location of the dam. ER 1110-8-2 defines the requirements for design floods to evaluate dam and spillway adequacy. Requirements for flood development and application are also provided.

7-2. Flood Hydrograph Analysis

a. Unit hydrograph method. The standard Corps procedure for computing flood hydrographs from catchments is the unit hydrograph method. The fundamental components are listed below:

(1) Analysis of rainfall and/or snowmelt to determine the time-distributed average precipitation input to each catchment area.

(2) Infiltration, or loss, analysis to determine the precipitation excess available for surface runoff.

(3) Unit hydrograph transforms to estimate the surface flow hydrograph at the catchment outflow location.

(4) Baseflow estimation to determine the subsurface contribution to the total runoff hydrograph.

(5) Hydrograph routing and combining to move catchment hydrographs through the basin and determine total runoff at desired locations.

For urban catchments, the kinematic-wave approach is often used to compute the surface flow hydrograph, instead of unit hydrograph transforms. Each of the standard flood runoff and routing procedures is described in Part 2 of EM 1110-2-1417. HEC-1 Flood *Hydrograph Package* (HEC 1990c) is a generalized computer program providing the standard methods for performing the required components for basin modeling.

b. Rainfall-runoff parameters. Whenever possible unit hydrographs and loss rate characteristics should be derived from the reconstitution of observed storm and flood events on the study watershed, or nearby watersheds with similar characteristics. The HEC-1 program has optimization routines to facilitate the determination of best-fit rainfall-runoff parameters for each event. When runoff records are not available at or near the location of interest, unit hydrograph and loss characteristics must be determined from regional studies of such characteristics observed at gauged locations. Runoff and loss coefficients can be related to drainage basin characteristics by multiple correlation analysis and mapping procedures, as described in Chapter 16, "Ungauged Basin Analysis" of EM 1110-2-1417.

c. Developing basin models. Flood hydrographs may be developed for a number of purposes. Basin models are developed to provide hydrographs for historical events at required locations where gauged data are not available. Even in large basins, there will be limited gauged data and many locations where data are desired. With some gauged data, a basin model can be developed and calibrated for observed flood events. Chapter 13 of EM 1110-2-1417 provides information on model development and calibration.

d. Estimating runoff. Basin models can estimate the runoff response under changing conditions. Even with historical flow records, many reservoir studies will require estimates of flood runoff under future, changed conditions. The future runoff with developments in the catchment and modifications in the channel system can be modeled with a basin runoff model.

e. Application. For reservoir studies, the most frequent application of flood hydrograph analysis is to develop hypothetical (or synthetic) floods. The three common applications are frequency-storms, SPF and PMF. Frequency-based design floods are used to develop flood-frequency information, like that required to compute expected annual flood damage. SPF and PMF are used as design standards to evaluate project performance under the more rare flood events.

7-3. Hypothetical Floods

a. General. Hypothetical floods are usually used in the planning and design of reservoir projects as a primary basis of design for some project features and to substantiate the estimates of extreme flood-peak frequency. Where runoff data are not available for computing frequency curves of peak discharge, hypothetical floods can be used to establish flood magnitudes for a specified frequency from rainstorm events of that frequency. This approach is not accurate where variations in soil-moisture conditions and rainfall distribution characteristics greatly influence flood magnitudes. In general, measured data should be used to the maximum extent possible, and when approximate methods are used, several approaches should be taken to compute flood magnitudes.

   b. Frequency-based design floods. In areas where infiltration losses are small, it may be feasible to compute hypothetical floods from rainfall amounts of a specified frequency and to assign that frequency to the flood event. NOAA publishes generalized rainfall criteria for the United States. They contain maps with isopluvial lines of point precipitation for various frequencies and durations. These point values are then adjusted for application to areas greater that 10 square miles, based on precipitation duration and catchment area. Section 13-4 of EM 1110-2-1417 provides information on simulation with frequency-based design storms.

   c. Standard project flood. The SPF is the flood that can be expected from the most severe combination of meteorologic and hydrologic conditions that are considered reasonably characteristic of the region in which the study basin is located. The SPF, which provides a performance standard for potential major floods, is based on the Standard Project Storm (SPS).

   (1) The SPS is usually an envelope of all or almost all of the storms that have occurred in a given region. The size of this storm is derived by drawing isohyetal maps of the largest historical storms and developing a depth-area curve for the area of maximum precipitation for each storm. Depth-area curves for storm rainfall of specified durations are derived from this storm-total curve by a study of the average time distribution of precipitation at stations representing various area sizes at the storm center. When such depth-area curves are obtained for all large storms in the region, the maximum values for each area size and duration are used to form a single set of depth-area-duration curves representing standard project storm hyetographs for selected area sizes, using a typical time distribution observed in major storms. EM 1110-2-1411 provides generalized SPS estimates for small and large drainage basins, and projects for which SPF estimates are required. The generalized rainfall criteria and recommended procedures for SPS computations for U.S. drainage basins located east of the 105th longitude are presented.

   (2) The SPF is ordinarily computed using the unit hydrograph approach with the SPS precipitation. The unit hydrograph and basin losses should be based on reasonable

values for a flood of this magnitude. Part 2 of EM 1110-2-1417 provides detailed information on the unit hydrograph procedure and the simulation of hypothetical floods is described in Chapter 13. The computer program HEC-1 *Flood Hydrograph Package* provides the SPS and SPF computation procedures, as described in the SPF determination manual.

   (3) While the frequency of the standard project flood cannot be specified, it can be used as a guide in extrapolating frequency curves because it is considered to lie within a reasonable range of rare recurrence intervals, such as between once in 200 years and once in 1,000 years.

   d. Probable maximum flood. The PMF is the flood that may be expected from the most severe combination of critical meteorologic and hydrologic conditions that are reasonably possible in the region. The PMF is calculated from the Probable Maximum Precipitation (PMP). The PMP values encompass the maximized intensity-duration values obtained from storms of a single type. Storm type and variations of precipitation are considered with respect to location, areal coverage of a watershed, and storm duration. The probable maximum storm amounts are determined in much the same way as are SPS amounts, except that precipitation amounts are first increased to correspond to maximum meteorologic factors such as wind speed and maximum moisture content of the atmosphere.

   (1) Estimates of PMP are based generally on the results of the analyses of observed storms. More than 600 storms throughout the United States have been analyzed in a uniform manner, and summary sheets have been distributed to government agencies and the engineering profession. These summary sheets include depth-area-duration data for each storm analyzed along with broad outlines of storm magnitudes and their seasonal and geographical variations. NWS (1977) Hydrometeorlogical Report No. 51 (HMR 51) contains generalized all-season estimates for the United States, east of the 105th longitude. The PMP is distributed in space and time to develop the PMS, which is a hypothetical storm that produces the PMF for a particular drainage basin.

   (2) NWS (1981) HMR No. 52 provides criteria and instructions for configuring the storm to produce the PMF. The precipitation on a basin is affected by the storm placement, storm-area size, and storm orientation. The HMR52 PMS (HEC 1984) computer program uses a procedure to produce maximum precipitation on the basin. However, several trials are suggested to ensure that the maximum storm is produced. The PMS is then input to a rainfall-runoff model to determine the flood runoff.

(3) The HMR52 User's Manual shows an example application with the HEC-1 Flood Hydrograph Package. The storm hyetographs can be written to an output file, in HEC-1 input format, or to an HEC-DSS file. HEC-1 can read the DSS file to obtain the basin precipitation.

(4) Hydrometeorlogical criteria are being updated for various areas of the country. A check should be made for the most recent criteria. Figure 13-3 in EM 1110-2-1417 shows the regional reports available in 1993. The HMR52 computer program does not apply to U.S. regions west of the 105th meridian.

(5) In the determination of both the SPF and the PMF, selection of rainfall loss rates and the starting storage of upstream reservoirs should be based on appropriate assumptions for antecedent precipitation and runoff for the season of the storm. Also, PMF studies should consider the capability of upstream reservoir projects to safely handle the PMF contribution from that portion of the watershed. There could be deficiencies in an upstream project spillway that significantly affects the downstream project's performance.

e. Storm duration. Hypothetical storms to be used for any particular category of hypothetical flood computation must be based on data observed within a region. For application in the design of local flood protection projects, only peak flows and runoff volumes for short durations are usually important. Accordingly, the maximum pertinent duration of storm rainfall is only on the order of the time of travel for flows from the headwaters to the location concerned. After a reasonable maximum duration of interest is established, rainfall amounts for this duration and for all important shorter durations must be established. For standard project storm determinations, this would consist of the amounts of observed rainfall in the most severe storms within the region that correspond to area sizes equal to the drainage area above the project. In the case of hypothetical storms and floods of a specified frequency, these rainfall amounts would correspond to amounts observed to occur with the specified frequency at stations spread over an area the size of the project drainage area. Larger rates and smaller amounts of precipitation would occur for shorter durations, as compared with the longer durations of interest. Once a depth-duration curve is established that represents the desired hypothetical storm rainfall, a time pattern must be selected that is reasonably representative of observed storm sequences. The HEC-1 computer program has the capability of accepting any depth-duration relation and selecting a reasonable time sequence. It is also capable of accepting specified time sequences for hypothetical storms.

f. Snowmelt contribution. Satisfactory criteria and procedures have not yet been developed for the computation of standard project and probable maximum snowmelt floods. The problem is complicated in that deep snowpack tends to inhibit rapid rates of runoff, and consequently, probable maximum snowmelt flood potential does not necessarily correspond to maximum snowpack depth or water equivalent. Snowpack and snowmelt differ at various elevations, thus adding to the complexity of the problem.

(1) Where critical durations for project design are short, high temperatures occurring with moderate snowpack depths after some melting has occurred will probably produce the most critical runoff. Where critical durations are long, as is the more usual case in the control of snowmelt floods, prolonged periods of high temperature or warm rainfall occurring with heavy snowpack amounts will produce critical conditions.

(2) The general procedure for the computation of hypothetical snowmelt floods is to specify an initial snowpack for the season that would be critical. In the case of SPF's a maximum observed snowpack should be assumed. The temperature sequence for SPF computation would be that which produces the most critical runoff conditions and should be selected from an observed historical sequence. In the case of PMF computation, the most critical snowpack possible should be used and it should be considerably larger or more critical than the standard project snowpack. The temperature pattern should be selected from historical temperature sequences augmented to represent probable maximum temperature for the season. Where simultaneous contribution from rainfall is possible, a maximum rainfall for the season should be added during the time of maximum snowmelt. This would require some moderation of temperatures to ensure that they are consistent with precipitation conditions. EM 1110-2-1406 covers snowmelt for design floods, standard project and maximum probable snowmelt flood derivation.

(3) Snowmelt computations can be made in accordance with an energy budget computation, accounting for radiation, evaporation, conductivity, and other factors, or by a simple relation with air temperature, which reflects most of these other influences. The latter procedure is usually more satisfactory in practical situations. Snowmelt, loss rate, and unit hydrograph computations can be made by using a computer program like Flood Hydrograph Analysis, HEC-1. EM 1110-2-1417 has detailed descriptions of each computational component.

Chapter 8
Water Surface Profiles

8-1. Introduction

a. General. Water surface profiles are required for most reservoir projects, both upstream and downstream from the project. Profile computations upstream from the project define the "backwater" effect due to high reservoir pool levels. The determination of real estate requirements are based on these backwater profiles. Water surface profiles are required downstream to determine channel capacity, flow depths and velocities, and other hydraulic information for evaluation of pre- and post-project conditions.

b. Choosing a method. The choice of an appropriate method for computing profiles depends upon the following characteristics: the river reach, the type of flow hydrograph, and the study objectives. The gradually varied, steady flow profile computation (e.g., HEC-2), is used for many studies. However, the selection of the appropriate method is part of the engineering analysis. EM 1110-2-1416 provides information on formulating a hydraulic study and a discussion of the analytical methods in general use. The following sections provide general guidance on the methods and the potential application in reservoir related studies.

8-2. Steady Flow Analysis

a. Method assumptions. A primary consideration in one-dimensional, gradually varied, steady flow analysis is that flow is assumed to be constant, in time, for the profile computation. Additionally, all the one-dimensional methods require the modeler to define the flow path when defining the cross-sectional data perpendicular to the flow. The basic assumptions of the method are as follows:

(1) Steady flow - depth and velocity at a given location do not vary with time.

(2) Gradually varied flow - depth and velocity change gradually along the length of the water course.

(3) One-dimensional flow - variation of flow characteristics, other than in the direction of the main axis of flow may be neglected, and a single elevation represents the water surface of a cross section perpendicular to the flow.

(4) Channel slope less than 0.1 m/m - because the hydrostatic pressure distribution is computed from the depth of water measure vertically.

(5) Averaged friction slope - the friction loss between cross sections can be estimated by the product of the representative slope and reach length.

(6) Rigid boundary - the flow cross section does not change shape during the flood.

b. Gradually varied steady flow. The assumption of gradually varied steady flow for general rainfall and snowmelt floods is generally acceptable. Discharge changes slowly with time and the use of the peak discharge for the steady flow computations can provide a reasonable estimate for the flood profile. Backwater profiles, upstream from a reservoir, are routinely modeled using steady flow profile calculations. However, inflow hydrographs from short duration, high intensity storms, e.g., thunderstorms, may not be adequately modeled assuming steady flow.

c. Downstream profile. Obviously, the downstream profile for a constant reservoir release meets the steady flow condition. Again, the consideration is how rapidly flow changes with time. Hydropower releases for a peaking operation may not be reasonably modeled using steady flow because releases can change from near zero to turbine capacity, and back, in a short time (e.g., minutes) relative to the travel time of the resulting disturbances. Dam-break flood routing is another example of rapidly changing flow which is better modeled with an unsteady flow method.

d. Flat stream profiles. Another consideration is calculating profiles for very flat streams. When the stream slope is less than 0.0004 m/m (2 ft/mile), there can be a significant loop in the downstream stage-discharge relationship. Also, the backwater effects from downstream tributaries, or storage, or flow dynamics may strongly attenuate flow. For slopes greater than 0.0009 m/m (5 ft/mile), steady flow analysis is usually adequate.

e. Further information. Chapter 6 of EM 1110-2-1416 *River Hydraulics* provides a detailed review of the assumptions of the steady flow method, data requirements, and model calibration and application. Appendix D provides information on the definition of river geometry and energy loss coefficients, which is applicable to all the one-dimensional methods.

8-3. Unsteady Flow Analysis

a. Unsteady flow methods. One-dimensional unsteady flow methods require the same assumptions listed in 8-2(a), herein, except flow, depth, and velocity can vary with time. Therefore, the primary reason for using unsteady flow methods is to consider the time varying nature of the problem. Examples of previously mentioned rapidly changing flow are thunderstorm floods, hydroelectric peaking operations, and dam-break floods. The second application of unsteady flow analysis consideration, mentioned above, is streams with very flat slopes.

b. Predicting downstream stages. Another application of unsteady flow is in the prediction of downstream stages in river-reservoir systems with tributaries, or lock-and-dam operations where the downstream operations affect the upstream stage. Flow may not be changing rapidly with time, but the downstream changes cause a time varying downstream boundary condition that can affect the upstream stage. Steady flow assumes a unique stage-discharge boundary condition that is stable in time.

c. Further information. Chapter 5, "Unsteady Flow," in EM 1110-2-1416 provides a detailed review of model application including selection of method, data requirements, boundary conditions, calibration, and application.

8-4. Multidimensional Analysis

a. Two- and three-dimensional modeling. Multidimensional analysis includes both two- and three-dimensional modeling. In river applications, two-dimensional modeling is usually depth-averaged. That is, variables like velocity do not vary with depth, so an average value is computed. For deep reservoirs, the variation of parameters with depth is often important (see Chapter 12, EM 1110-2-1201). Two-dimensional models, for deep reservoirs, are usually laterally-averaged. Three-dimensional models are available; however, their applications have mostly been in estuaries where both the lateral and vertical variation are important.

b. Two-dimensional analysis. Two-dimensional, depth-averaged analysis is usually performed in limited portions of a study area at the design stage of a project. The typical river-reservoir application requires both the direction and magnitude of velocities. Potential model applications include areas upstream and downstream from reservoir outlets. Additionally, flow around islands, and other obstructions, may require two-dimensional modeling for more detailed design data.

c. Further information. Chapter 4 of EM 1110-2-1416 provides a review of model assumptions and typical applications.

8-5. Movable-Boundary Profile Analysis

a. Reservoirs. Reservoirs disrupt the flow of sediment when they store or slow down water. At the upper limit of the reservoir, the velocity of inflowing water decreases and the ability to transport sediment decreases and deposition occurs. Chapter 9 herein presents reservoir sediment analysis. Reservoir releases may be sediment deficient, which can lead to channel degradation downstream from the project because the sediment is removed from the channel.

b. River and reservoir sedimentation. EM 1110-2-4000 is the primary Corps reference on reservoir sedimentation. Chapter 3 covers sediment yield and includes methods based on measurement and mathematical models. Chapter 4 covers river sedimentation, and Chapter 5 presents reservoir sedimentation. Section III, of Chapter 5, provides an overview of points of caution, sedimentation problems associated with reservoirs, and the impact of reservoirs on the stream system. Section IV provides information on levels of studies and study methods.

b. Further information. Chapter 7 of EM 1110-2-1416 presents water surface profile computation with movable boundaries. The theory, data requirements and sources, plus model development and application are all covered. The primary math models, HEC-6 *Scour and Deposition in Rivers and Reservoirs* (HEC 1993) and *Open-Channel Flow and Sedimentation* TABS-2 (Thomas and McAnally 1985) two-dimensional modeling package, are also described. The focus for the material is riverine.

Chapter 9
Reservoir Sediment Analysis

9-1. Introduction

a. Parameters of a natural river. Nature maintains a very delicate balance between the water flowing in a natural river, the sediment load moving with the water, and the stream's boundary. Any activity which changes any one of the following parameters:

- water yield from the watershed.

- sediment yield from the watershed.

- water discharge duration curve.

- depth, velocity, slope or width of the flow.

- size of sediment particles.

or which tends to fix the location of a river channel on its floodplain and thus constrains the natural tendency will upset the natural trend and initiate the formation of a new one. The objective of most sediment studies is to evaluate the impact on the flow system resulting from changing any of these parameters.

b. Changes caused by reservoirs. Reservoirs interrupt the flow of water and, therefore, sediment. In terms of the above parameters, the reservoir causes a change in the upstream hydraulics of flow depth, velocity, etc. by forcing the energy gradient to approach zero. This results in a loss of transport capacity with the resulting sediment deposition in the reservoir. The reservoir also alters the downstream water discharge-duration relation and reduces the sediment supply which may lead to the degradation of the downstream channel.

c. Areas of analysis. Sedimentation investigations usually involve the evaluation of the existing condition as well as the modified condition. The primary areas of reservoir sediment analysis are the estimation of volume and location of sediment deposits in the reservoir and the evaluation of reservoir releases' impact on the downstream channel system. Sediment deposits start in the backwater area of the reservoir, which increase the elevation of the bed profile and the resulting water surface profile. However, reservoirs may also cause sediment deposits upstream from the project, which affect the upstream water surface profiles.

d. Further information. The primary Corps reference for sediment analysis is EM 1110-2-4000. Major topics include developing a study work plan, sediment yield, river sedimentation, reservoir sedimentation, and model studies.

9-2. Sediment Yield Studies

a. General. Sediment yield studies determine the amount of sediment that leaves a basin for an event or over a period of time. Sediment yield, therefore, involves erosion processes as well as sediment deposition and delivery to the study area. The yield provides the necessary input to determine sedimentation impacts on a reservoir.

b. Required analysis. Each reservoir project requires a sediment yield analysis to determine the storage depletion resulting from the deposition of sediment during the life of the project. For most storage projects, as opposed to sediment detention structures, the majority of the delivered sediment is suspended. However, the data required for the headwater reaches of the reservoir should include total sediment yield by particle size because that is where the sands and gravels will deposit.

c. Further information. Corps of Engineer methods for predicting sediment yields are presented in Appendix C of EM 1110-2-4000. A literature review, conducted by the Hydrologic Engineering Center under the Land Surface Erosion research work unit, showed numerous mathematical models are available to estimate sediment discharge rates from a watershed and the redistribution of soil within a watershed. An ETL on the methods will be issued soon.

9-3. Reservoir Sedimentation Problems

a. Sediment deposition. As mentioned above, the primary reservoir sediment problem is the deposition of sediment in the reservoir. The determination of the sediment accumulation over the life of the project is the basis for the sediment reserve. Typical storage diagrams of reservoirs, showing sediment (or dead) storage at the bottom of the pool can be misleading. While the reservoir storage capacity may ultimately fill with sediment, the distribution of the deposits can be a significant concern during the life of the project. The reservoir sedimentation study should forecast sediment accumulation and distribution over the life of the project. Sediment deposits in the backwater area of the reservoir may form deltas, particularly in shallow reservoirs. A number of problems associated with delta formations are discussed below.

(1) Deposits forming the delta may raise the water surface elevation during flood flows, thus requiring special

consideration for land acquisition. In deep reservoirs, this is usually not a problem with the reservoir area because project purposes dictate land acquisitions or easements. Deltas tend to develop in the upstream direction. In shallow reservoirs, the increase in water surface elevation is a problem even within the reservoir area. That is, floods of equal frequency may have higher water surface elevations after a project begins to develop a delta deposit than was experienced before the project was constructed. Land acquisition studies must consider such a possibility.

(2) Aggradation problems are often more severe on tributaries than on the main stem. Analysis is complicated by the amount of hydrologic data available on the tributaries, which is usually less than on the main stem itself. Land use along the tributary often includes recreation sites, where aggradation problems are particularly undesirable.

(3) Reservoir deltas often attract phreatophytes due to the high moisture level. This may cause water-use problems due to their high transpiration rate.

(4) Reservoir delta deposits are often aesthetically undesirable.

(5) Reservoir sediment deposits may increase the water surface elevation sufficiently to impact on the groundwater table, particularly in shallow impoundments.

(6) In many existing reservoirs, the delta and backwater-swamp areas support wildlife. Because the characteristics of the area are closely controlled by the operation policy of the reservoir, any reallocation of storage would need to consider the impact on the present delta and swamp areas.

b. Upstream projects. It is important to identify and locate all existing reservoirs in a basin where a sediment study is to be made. The projects upstream from the point of analysis potentially modify both the sediment yield and the water discharge duration curve. The date of impoundment is important so that observed inflowing sediment loads may be coordinated with whatever conditions existed in the basin during the periods selected for calibration and verification. Also, useful information on the density of sediment deposits and the gradation of sediment deposits along with sediment yield are often available from other reservoirs in the basin. Information on the rate of sediment deposition that has occurred at other reservoir sites in the region is the most valuable information when estimating sediment deposition for a new reservoir.

9-4. Downstream Sediment Problems

a. Channel degradation. Channel degradation usually occurs downstream from the dam. Initially, after reservoir construction, the hydraulics of flow (velocity, slope, depth, and width) remain unchanged from pre-project conditions. However, the reservoir acts as a sink and traps sediment, especially the bed material load. This reduction in sediment delivery to the downstream channel causes the energy in the flow to be out of balance with the boundary material for the downstream channel. Because of the available energy, the water attempts to re-establish the former balance with sediment load from material in the stream bed, and this results in a degradation trend. Initially, degradation may persist for only a short distance downstream from the dam because the equilibrium sediment load is soon re-established by removing material from the stream bed.

b. Downstream migratory degradation. As time passes, degradation tends to migrate downstream. However, several factors are working together to establish a new equilibrium condition in this movable-boundary flow system. The potential energy gradient is decreasing because the degradation migrates in an upstream-to-downstream direction. As a result, the bed material is becoming coarser and, consequently, more resistant to being moved. This tendency in the main channel has the opposite effect on tributaries. Their potential energy gradient is increasing which results in an increase in transport capacity. This will usually increase sediment passing into the main stem which tends to stabilize the main channel resulting in less degradation than might be anticipated. Finally, a new balance will tend to be established between the flowing water-sediment mixture and the boundary.

c. Extent of degradation. The extent of degradation is complicated by the fact that the reservoir also changes the discharge duration curve. This will impact for a considerable distance downstream from the project because the existing river channel reflects the historical phasing between flood flows on the main stem and those from tributaries. That phasing will be changed by the operation of the reservoir. Also, the reduced flow will probably promote vegetation growth at a lower elevation in the channel. The result is a condition conducive to deposition in the vegetation. Detailed simulation studies should be performed to determine future channel capacities and to identify problem areas of excessive aggradation or degradation. All major tributaries should be included.

9-5. Sediment Water Quality

a. Sediments and pollutants. When a river carrying sediments and associated pollutants enters a reservoir, the flow velocity decreases and the suspended and bed load sediments start settling down. Reservoirs generally act as depositories for the sediments because of their high sediment trap efficiency. Due to a high adsorption capacity, sediments act as sinks for contaminants in the reservoirs and, in agricultural and industrial areas, may contain PCB's, chlorinated hydrocarbon pesticides, oil and grease, heavy metals, coliform bacteria, or mutagenic substances. Burial of these contaminants by sedimentation may be an important factor and an effective process in isolating potentially toxic substances from surface waters and important biological populations. Toxic inorganic and organic contaminants associated with the sediments can also be bioconcentrated by the aquatic organisms present in reservoirs.

b. Monitoring chemical contaminants. These incoming sediments and associated pollutants significantly affect the water quality of the reservoir pool and downstream releases. Therefore, it is essential that these sediment reservoir interactions be characterized by their depositional behavior, particle size distribution, and pollutant concentrations to successfully plan a management strategy to quantify contaminant movement within reservoirs. Analytical and predictive methods to assess the influence of contaminated sediments in reservoirs have not been developed enough to be used in Corps field offices, but WES Instruction Report E-86-1, "General Guidelines for Monitoring Contaminants in Reservoirs" (Waide 1986), does provide general guidance on the design and conduct of programs for monitoring chemical contaminants in reservoir waters, sediments, and biota.

c. Sedimentation patterns. Sedimentation patterns can often be associated with water quality characteristics. There seems to be a relationship between longitudinal gradients in water quality (a characteristic of many reservoirs) and sediment transport and deposition. High concentrations of inorganic particulates can reduce light availability near inflows and thus influence algal production and decrease dissolved oxygen. The association of dissolved substances, such as phosphorus, with suspended solids may act to reduce or buffer dissolved concentrations, thus influencing nutrient availability.

9-6. Sediment Investigations

a. General. The level of detail required for the analysis of any sediment problem depends on the objective of the study. Chapter 1 of EM 1110-2-4000 describes staged sedimentation studies in Section I. Section II, of that chapter, provides reporting requirements. Problem identification and the development of a study work plan are covered in Chapter 2.

b. Sediment deposits. Considering a dam site as an important natural resource, it is essential to provide enough volume in the reservoir to contain anticipated deposits during the project life. If the objective of a sediment study is just to know the volume of deposits for use in screening studies, then trap efficiency techniques can provide a satisfactory solution. The important information that must be available is the water and sediment yields from the watershed and the capacity of the reservoir. Chapter 3 of EM 1110-2-4000 covers sediment yield. Section 3-7 provides information on reservoir sedimentation, including trap efficiency.

c. Land acquisition. If the sediment study must address land acquisition for the reservoir, then knowing only the volume of deposits is not sufficient. The location of deposits must also be known, and the study must take into account sediment movement. This generally requires simulation of flow in a mobile boundary channel. Sorting of grain sizes must be considered because the coarser material will deposit first, and armoring must be considered because scour is involved. Movable-bed modeling is useful to predict erosion or scour trends downstream from the dam, general aggradation or degradation trends in river channels, and the ability of a stream to transport the bed-material load. The computer program, HEC-6 *Scour and Deposition in Rivers and Reservoirs* (HEC 1993), is designed to provide long-term trends associated with changes in the frequency and duration of the water discharge and/or stage or from modifying the channel geometry.

d. Details of investigations. The details of reservoir sedimentation investigations are covered in Chapter 5 of EM 1110-2-4000. The primary emphasis is on the evaluation of the modified condition, which includes consideration of quality and environmental issues. The levels of sedimentation studies and methods of analysis are presented in Section IV of Chapter 5. Model studies and a short review of HEC-6 and the two-dimensional TABS-2 modeling system are covered in Chapter 6.

PART 3

RESERVOIR STORAGE REQUIREMENTS

Chapter 10
Flood-Control Storage

10-1. General Considerations

a. Reservoir flood storage. Where flood damage at a number of locations on a river can be significantly reduced by construction of one or more reservoirs, or where a reservoir site immediately upstream from one damage center provides more economical protection than local protection works, reservoir flood storage should be considered. Whenever such reservoirs can serve needs other than flood control, the integrated design and operation of the project for multipurpose use should be considered.

b. Flood-control features. In planning and designing the flood-control features of a reservoir, it is important that the degree and extent of continuous ensured protection be no less than that provided by a local protection project, if the alternatives of reservoir construction or channel and levee improvement are to be evaluated fairly. This means that the storage space and release schedule for flood control must be provided at all times when the flooding potential exists. In some regions this may be for the entire year, but more commonly there are dry seasons when the flood potential is greatly reduced and storage reservation for flood control can be reduced correspondingly. Except where spring snowmelt floods can be forecasted reliably or where safe release rates are sufficient to empty flood space in a very short time, it is not ordinarily feasible to provide flood-control space only after a flood is forecasted. Space must be provided at all times during the flood season unless it can be demonstrated that the necessary space can be evacuated on a realistic forecast basis. Also, space may be reduced if less storage is needed due to low snowpack, or there is some other reliable basis for long range flood forecasting.

c. Runoff volume durations. Whereas the peak rates of runoff are critical in the design of local protection projects, runoff volumes for pertinent durations are critical in the design of reservoirs for flood control. The critical durations will be a function of the degree of flood protection selected and of the release rate or maximum rate of flow at the key downstream control point. As the proposed degree of protection is increased and as the proposed rates of controlled flows at key damage centers are reduced, the critical duration is increased. If this critical duration corresponds to the duration of a single rainstorm period or a single snowmelt event, the computation of hypothetical floods from rainfall and snowmelt can constitute the principle hydrologic design element. On the other hand, if the critical duration is much longer, hypothetical floods and sequences of hypothetical floods computed from rainfall or snowmelt become less dependable as guides to design. It then is necessary to base the design primarily on the frequency of observed runoff volumes for long durations. Even when this is done, it will be advisable to construct a typical hydrograph that corresponds to runoff volumes for the critical duration and that reasonably characterizes hydrographs at the location, in order to examine the operation of the proposed project under realistic conditions.

d. Hypothetical flood simulations. When hypothetical floods are selected, they must be routed through the proposed reservoir under the operation rules that would be specified for that particular design. In effect, a simulation study of the proposed project and operation scheme would be conducted for each flood. It is also wise to simulate the operation for major floods of historical record in order to ensure that some peculiar feature of a particular flood does not upset the plan of operation. With present software, it is relatively inexpensive to perform a complete period of record simulation once the flood-control storage is set.

10-2. Regulated Release Rates

a. Flood reduction purposes. For flood reduction purposes reservoirs must store only the water that cannot be released without causing major damage downstream. If more water can be released during a flood, less water needs to be stored. Thus, less storage space needs to be planned for flood control. Because reservoir space is costly and usually in high demand for other purposes, good flood-control practice consists of releasing water whenever necessary at the highest practical rates so that a minimum amount of space need be reserved for flood control. As these rates increase, it becomes costly also to improve downstream channels and to provide adequate reservoir outlets, so there is an economic balance between release rates and storage capacity for flood control. In general, it is economical to utilize the full nondamage capacity of downstream channels, and it may pay to provide some additional channel or levee improvements downstream. However, as described in paragraph *f*, full channel capacity may not be available, so analyses should consider the impact of reduced capacity.

b. Channel capacities. Channel capacities should be evaluated by examing water-surface profile data from actual flood events whenever possible. Under natural channel conditions, it will ordinarily be found that floods which occur more frequently than once in two years are not seriously damaging, while larger floods are.

c. Minor versus major damage releases. In some cases, it is most economical to sustain minor damage by releasing flows above nondamaging stages in order to accommodate major floods and thereby protect the more important potential damage areas from flooding. In such situations, a stepped-release schedule designed to protect all areas against frequent minor floods, with provision to increase releases after a specified reservoir stage is reached, might be considered. However, such a plan has serious drawbacks in practice because protection of the minor damage areas would result in greater improvements in those areas; and it soon becomes highly objectionable, if not almost impossible, to make the larger releases when they are required for protection of major damage areas. In any case, it is necessary to make sure that the minor damage areas are not flooded more frequently or severely with the project than they would have been without it.

d. Maintenance and zoning. It is important on all streams in developed areas to provide for proper maintenance of channel capacity and zoning of the floodplain where appropriate. This is vital where upstream reservoirs are operated for flood control because proper reservoir regulation depends as much on the ability to release without damage as it does on the ability to store. Minor inadequacies in channel capacity can lead to the loss of control and result in major flooding. This situation is aggravated because the reduced frequency of flooding below reservoirs and the ability to reduce reservoir releases when necessary often increase the incentive to develop the floodplain and sometimes even remove the incentive for maintaining channel capacity.

e. Forecasted runoff. When a reservoir is located some distance upstream from a damage center, allowance must be made for any runoff that will occur in the intermediate area. This runoff must be forecasted, a possible forecast error added, and the resulting quantities subtracted from project channel capacity to determine per-missible release rates considering attenuation when routing the release from the dam to the damage center and the contribution of flow from the intermediate drainage area. Also, with high intensity rainfall, the added rainfall depth to the total downstream channel flow should be considered.

f. Delaying flood releases. Experience in the flood-control operation of reservoirs has demonstrated that the actual operation does not make 100 percent use of downstream channel capacities. Due to many contributing factors average outflows during floods are less than maximum permissible values. It is usually wise to approach maximum release rates with caution, in order to ascertain any changes in channel capacity that have taken place since the last flood, and this practice reduces operational efficiency. It may be necessary to delay flood releases to permit removal of equipment, cattle, etc., from areas that would be flooded. Releases might be curtailed temporarily in order to permit emergency repairs to canals, bridges, and other structures downstream. If levees fail, releases might be reduced in order to hasten the drainage of flooded areas. Release can be reduced in order to facilitate rescue operations. These and various other conditions result in reduced operation efficiency during floods. To account for this, less nondamage flow capacity than actually exists (often about 80 percent) is assumed for design studies. It is important, however, that every effort be made in actual operation to effect the full non-damage releases in order to attain maximum flood-control benefits.

g. Gradually increasing and decreasing releases. During flood operations, reservoir releases must be increased and decreased gradually in order to prevent damage and undue hardship downstream. Gradually increasing releases will usually permit an orderly evacuation of people, livestock, and equipment from the river areas downstream. If releases are curtailed too rapidly, there is some danger that the saturated riverbanks will slough and result in the loss of valuable land or damage to levees.

10-3. Flood Volume Frequencies

a. Critical durations. Flood volume frequency studies usually consist of deriving frequency curves of annual maximum volumes for each of various specified durations that might be critical in project design. Critical durations range from a few hours in the case of regulating "cloudburst" floods to a few months where large storage and very low release rates prevail. The annual maximum volumes for a specific duration are usually expressed as average rates of flow for that duration. It is essential that these flows represent a uniform condition of development for the entire period of observation, preferably unregulated conditions. Procedures for computing the individual frequency curves are discussed briefly in Chapter 6 herein and are described in detail in EM 1110-2-1415.

b. Flood-control space requirement. Determination of the flood-control space needed to provide a selected degree of protection is based on detailed hydrograph analysis, but a general evaluation can be made as illustrated in Figure 10-1. The curve of runoff versus duration is obtained from frequency studies of runoff volumes or from SPF studies at the location. The tangent line represents a uniform flow equal to the project release capacity (reduced by an appropriate contingency factor). The intercept represents the space required for control of the flood. The chart demonstrates that a reservoir capable of storing

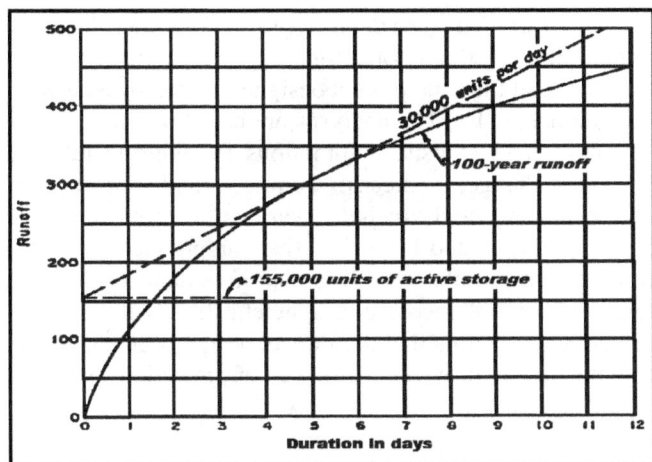

Figure 10-1. Flood-control space requirement

155,000 units of water and releasing 30,000 units per day can control 100-year runoff for any duration, and that the critical duration (period of increasing storage) is about 5 days. The volume-duration curve would be made for each damage area and should include more than 100 percent of the local uncontrolled runoff downstream from the reservoir and above the control point in order to allow for errors of forecast which would be reflected in reduced project releases. If this local runoff appreciably exceeds nondamage flow capacity at the damage centers, the volume over and above the flow capacity is damaging water that cannot be stored in the project reservoir.

10-4. Hypothetical Floods

a. Two classes. Two classes of hypothetical floods are important in the design of reservoirs for flood control. One is a balanced flood that corresponds to a specified frequency of occurrence; the other is a flood that represents a maximum potential for the location, such as the SPF or PMF. ER 1110-8-2(FR) sets forth hydrologic engineering requirements for selecting and accommodating inflow design floods for dams and reservoirs.

b. Specified frequencies. A hypothetical flood corresponding to a specified frequency should contain runoff volumes for all pertinent durations corresponding to that specified frequency. The derivation of frequency curves is as discussed in the preceding section. A balanced flood hydrograph is constructed by selecting a typical hydrograph pattern and adjusting the ordinates so that the maximum volumes for each selected duration correspond to the volumes for that duration at the specified frequency.

c. Longer duration floods. Where flood durations longer than the typical single-flood duration are important in the design, a sequence of flood hydrographs spaced reasonably in time should be used as a pattern flood. In order to represent average natural sequences of flood events, the largest portions of the pattern flood should ordinarily occur at or somewhat later than the midpoint of the entire pattern, because rainfall sequences are fairly random but ground conditions become increasingly wet and conducive to larger runoff as any flood sequence continues.

d. Maximum flood potential. Two types of hypothetical floods that represent maximum flood potential are important in the design of reservoirs. The PMF, which is the largest flood that is reasonably possible at the location, is ordinarily the design flood for the spillway of a structure where loss of life or major property damage would occur in the event of project failure. The SPF, which represents the largest flood for that location that is reasonably characteristic for the region, is a flood of considerably lesser magnitude and represents a high degree of design for projects protecting major urban and industrial areas. These floods can result from heavy rainfall or from snowmelt in combination with some rainfall.

e. Computing hydrographs. SPF and PMF hydrographs are computed from the storm hyetographs by unit hydrograph procedures. In the case of the SPF, ground conditions that are reasonably conducive to heavy runoff are used. In the case of the PMF, the most severe ground conditions that are reasonably consistent with storm magnitudes are used. A general description of these analyses is provided in Chapter 7 of this manual. Detailed methods for performing these computations are described in EM 1110-2-1417. The computer program HEC-1 *Flood Hydrograph Package* contains routines for computing floods from rainfall and snowmelt and also contains standard project criteria for the eastern United States.

10-5. Operation Constraints and Criteria

a. General. As stated earlier, whenever flood releases are required, it is imperative that they be made at maximum rates consistent with the conditions downstream. This means that the outlets should be designed to permit releases at maximum rates at all reservoir levels within the flood-control space. In some cases where controlled releases are very high, such an outlet design is not economical, and releases at lower stages might be restricted because of limited outlet capacity. This constraint, of course, should be taken into account during the design studies.

b. Downstream damage centers. Where damage centers are at some distance downstream from the reservoir, local runoff below the reservoir and above the damage center must be considered when determining releases to be made. This will ordinarily require some forecasting of the local runoff and, consequently, some estimate of the forecast uncertainty. The permissible release at any time is determined by adding a safe error allowance to the forecasted local inflow and subtracting this sum from the nondamaging flow capacity.

c. Rate-of-change of release. The rate-of-change of release must be restricted to the maximum changes that will not cause critical conditions downstream. As a practical matter, these rates-of-change of release should be less than the rates-of-change of flow that occurred before the reservoir was built. After the main flood has passed, water stored in the flood-control space must be released and maximum rates of release will continue until the desired amount of water is released, except that the rate of release should be decreased gradually toward the end of the release period. This reduction in release must be started while considerable flood waters remain in the reservoir in order that water retained for other purposes is not inadvertently released. Schedules for this operation are discussed in Part 3.

10-6. Storage Capacity Determinations

a. Determining required storage capacity. The storage capacity required to regulate a specific flood (represented by a flood hydrograph at the dam) to a specified control discharge immediately downstream of the dam is determined simply by routing the hydrograph through a hypothetical reservoir with unlimited storage capacity and noting the maximum storage. However, there are many special practical considerations that complicate this process. Release rates should not be changed suddenly; therefore, the routing should conform to criteria that specify the maximum rate of change of release. Also, outlet capacities might not be adequate to supply full regulated releases with low reservoir stages. If this is the case, a preliminary reservoir design is required in order to define the relation of storage capacity to outlet capacity.

b. Specified flood. In the more common cases, where damage centers exist at some distance downstream of the reservoir, the storage requirement for a specified flood is determined by successive approximations, operating the hypothetical reservoir to regulate flows at each damage center to nondamaging capacity, and allowing for local inflow and for some forecasting error.

c. Detailed operational study. Although there are approximate methods for estimating storage capacity, it is essential that the final project design be tested by a detailed operational study. The analyses are based on actual outlet capacities and realistic assumptions for limiting rates of release change, forecast errors, and operational contingencies, and include various combinations of reservoir inflow and local flow that can produce a specific downstream flood event. It is also important to route the largest floods of record and synthetic floods through the project to determine that the project design is adequate and that the project provides the degree of protection for which it was designed.

d. Seasonal distribution of storage requirements. Where some of the flood-control space will be made available for other uses during the dry season, a seasonal distribution of flood-control storage requirement should be developed. The most direct approach to this entails the construction of runoff frequency curves for each month of the year. The average frequency of the design flood during the rainy-season months can be used to select flood magnitudes for other months. These could then serve as a basis for determining the amount of space that must be made available during the other months.

e. Further information. Sequential routing in planning, design, and operation of flood-control reservoirs can be accomplished with the computer program HEC-5 *Simulation of Flood Control and Conservation Systems* (HEC 1982c).

10-7. Spillways

Spillways are provided to release floodwater which normally cannot be passed by other outlet works. The spillway is sized to ensure the passage of major floods without overtopping the dam. A general discussion of spillways is provided in Section 4-2 of EM 1110-2-3600. EM 1110-2-1603 describes the technical aspects of design for the hydraulic features of spillways and ER 1110-8-2(FR) sets forth requirements for selecting and accommodating inflow design floods.

a. Spillway design flood. The spillway design flood is usually selected as a large hypothetical flood derived from rainfall and snowmelt. Other methods of estimating extreme flood magnitudes, such as flood-frequency analysis, are not reliable due to limited observations. The selection of a spillway design flood depends on the policies of the construction agency and regulations governing dam construction. Usually, the spillways for major dams, whose failure might constitute a major disaster, are designed to pass the PMF without a major failure; however, the spillways for many small dams are designed for smaller floods such as the SPF.

b. Hydrologic design. The hydrologic design of a spillway is accomplished by first estimating a design and then testing it by routing the spillway design flood. In routing the spillway design flood, the initial reservoir stage should be as high as reasonably expected at the start of such a major flood, considering the manner in which the reservoir is planned to operate or how in the future the reservoir might operate differently from the planned operation. In the case of ungated spillways, it is possible that the outlets of the dam will be closed gradually as the spillway goes into operation, in order to delay damaging releases as long as possible and possibly to prevent them. However, if spillway flows continue to increase, it may be necessary to reopen the outlets. In doing so, care should be exercised to prevent releases from exceeding maximum inflow quantities. The exact manner in which outlets will be operated should be specified so that the spillway design will be adequate under conditions that will actually prevail after project construction. Consideration should be given to the possibility that some outlets or turbines might be out of service during flood periods.

c. Large spillway gates. The operation of large spillway gates can be extremely hazardous, since opening them inadvertently might cause major flooding at downstream areas. Their operation should be controlled by rigid regulations. In particular, the opening of the gates during floods should be scheduled on the basis of inflows and reservoir storage so that the lake level will continue to rise as the gates are opened. This will ensure that inflow exceeds outflow as outflows are increased. The adequacy of a spillway to pass the spillway design flood is tested for gated spillways in the same manner as for ungated spillways described above. Methods for developing spillway-gate operation regulations are described in Chapter 14.

d. Preventing overtopping. To ensure that the spillway is adequate to protect the structure from overtopping, some amount of freeboard is added to the dam above the maximum pool water-surface elevation. This can vary from zero for structures that can withstand overtopping to 2 m or more for structures where overtopping would constitute a major hazard. The freeboard allowance accounts for wind set and wave action. Methods for estimating these quantities are discussed in Chapter 15. Risk analysis should be performed to determine the appropriate top-of-dam elevation.

e. Spillway types. While the spillway is primarily intended to protect the structure from failure, the fact that it can cause some water to be stored above ordinary full pool level (surcharge storage) is of some consequence in reducing downstream flooding. Narrow, ungated spillways require higher dams and can, therefore, be highly effective in partially regulating floods that exceed project design magnitude, whereas wide spillways and gated spillways are less effective for regulating floods exceeding design magnitude. Where rare floods can cause great damage downstream, the selection of spillway type and characteristics can appreciably influence the benefits that are obtained for flood control. Accordingly, it is not necessarily the least costly spillway that yields the most economical plan of development. In evaluating flood-control benefits, computing frequency curves for regulated conditions should be based on spillway characteristics and operation criteria as well as on other project features.

10-8. Flood-Control System Formulation

a. Objectives. The objectives of system formulation are to identify the individual components, determine the size of each, determine the order in which the system components should be implemented, and develop and display the information required to justify the decisions and thus secure system implementation. Section 4-10 describes several formulation strategies.

b. Criteria. Criteria for system formulation are needed to distinguish the best system from among competing alternative systems. The definition of "best" is crucial. A reasonable viewpoint would seem to recognize that simply aggregating the most attractive individual components into a system, while assuring physical compatibility, could result in the inefficient use of resources because of system effects, data uncertainty, and the possibility that all components may not be implemented. It is proposed that the best system be considered to be as follows:

(1) The system that includes the obviously good components while preserving flexibility for modification of components at future dates.

(2) The system which could be implemented at a number of stages, if staging is possible, such that each stage could stand on its own merits (be of social value) if no more components were to be added.

c. General guidance. General guidance for formulation criteria are contained in the Principles and Standards (Water Resources Council 1973). The criterion of economic efficiency from the national viewpoint has been interpreted to require that each component in a system should be incrementally justified, that is, each component addition to a system should add to the value (net benefits) of the total system. The environmental quality criteria can be viewed as favoring alternatives that can be structured to minimize adverse environmental impacts and provide opportunities for mitigation measures. Additional criteria that are not as formally stated as U.S. national policy are important in decisions among alternatives. A formulated flood-control system must draw sufficient support from responsible authorities in order to be implemented. In addition, flood-control systems should be formulated so that a minimum standard of performance (degree of risk) is provided so that public safety and welfare are adequately protected.

d. Environmental and other assessments. Of these criteria, only the national economic efficiency and minimum performance standard have generally accepted methods available for their rigorous inclusion in formulation studies. Environmental quality analysis and social/political/institutional analyses related to implementation have not developed technology applicable on a broad scale. As a consequence, these criteria must guide the formulation studies but, as yet, probably cannot directly contribute in a structured formulation strategy. In discussions that follow, focus is of necessity upon the economic criteria with acceptable performance as a constraint, with the assumption that the remaining criteria will be incorporated when the formulation strategy has narrowed the range of alternatives to a limited number for which the environmental and other assessments can be performed.

e. Degrees of uncertainty. There will be varying degrees of uncertainty in the information used in system formulation. The hydrology will be better defined near gauging stations than it is in remote areas, and certain potential reservoirs will have been more thoroughly investigated than others. In addition, the accuracy of economic data, both costs and value, existing or projected, is generally lower than the more physically based data. Also, since conditions change over time, the data must be continuously updated at each decision point. The practical accommodation of information uncertainty is by limited sensitivity analysis and continuing reappraisal as each component of a system is studied for implementation.

f. Sensitivity analysis. Sensitivity analysis has, as its objective, the identification of either critical elements of data, or particularly sensitive system components, so that further studies can be directed toward firming up the uncertain elements or that adjustments in system formulation can be made to reduce the uncertainty. Because

combinations of historic and synthetic floods are typically used to evaluate reservoir flood-reduction performance (i.e., to develop regulated conditions frequency relations at damage index stations), particular attention must be paid to the selection or development of the system hydrology. The problem arises when evaluating complex reservoir systems with many reservoirs above common damage centers. The problem increases with the size and complexity of the basin because the storm magnitudes and locations can favor one reservoir location over another. There are a large number of storm centerings that could yield similar flows at a particular control point. Because of this, the contribution of a specific system component to reduced flooding at a downstream location is uncertain and dependent upon storm centering. This makes the selection or development of representative centerings crucial if all upstream components are to be evaluated on a comparable basis.

g. Desired evaluation. The desired evaluation for regulated conditions is the expected or average condition so that economic calculations are valid. The representative hydrograph procedure is where several proportions (ratios of one or more historic or synthetic events used to represent system hydrology) are compatible with the simulation technique used, but care must be taken to reasonably accommodate the storm centering uncertainty. Testing the sensitivity of the expected annual damage to the system hydrology (event centering) is appropriate and necessary. Even if all historical floods of record are used, there still may be some bias in computing expected annual damages if most historical floods were, by chance, centered over a certain part of the basin and not over others. For instance, one reservoir site may have experienced several severe historical floods, while another site immediately adjacent to the area may, due to chance, not have had any severe floods.

10-9. General Study Procedure

After various alternative locations are selected for a reservoir site to protect one or more damage centers, the following steps are suggested for conducting the required hydrologic engineering studies:

a. Obtain a detailed topographic map of the region showing the locations of the damage areas, of proposed reservoir sites, and of all pertinent precipitation, snowpack, and stream-gauging stations. Prepare a larger scale topographic map of the drainage basin tributary to the most downstream damage location. Locate damage centers, project sites, pertinent hydrologic measurement stations, and drainage boundaries above each damage center, project site, and stream-gauging station. Measure all pertinent tributary areas.

b. Establish stage-discharge relations for each damage reach, relating the stages for each reach to a selected index location in that reach; procedures for doing this are described in *Flood-Damage Analysis Package User's Manual* (HEC 1990b). Where local protection works are considered part of an overall plan of improvement, establish the stage-discharge relation for each plan of local protection.

c. Obtain area- and storage-elevation curves for each reservoir site; select alternative reservoir capacities as appropriate for each site; select outlet and spillway rating curves for each reservoir, and develop a plan of flood-control operation for each reservoir. Determine maximum regulated flows for each damage center.

d. Estimate the maximum critical duration of runoff for any of the plans of improvement, considering the relation of regulated flows at damage centers to unregulated flood hydrographs of design magnitude at those damage centers. Prepare frequency curves of unregulated peak flows and volumes of each of various representative durations, as described for peak flows in Chapter 6, for each damage center index location, and for each reservoir site. If seasonal variation of flood-control space is to be considered, these curves should be developed for each season.

e. The two basic approaches for flood-control simulation are complete period-of-record analysis and representative floods analysis. If flooding can occur during any time of the year, the complete sequential analysis might be favored. However, if there is a separable flood season, e.g., in the western states, then the representative storm approach may be sufficient. For the storm approach, develop data for historical floods with storm centerings throughout the basin and use several proportions of those floods to obtain flows at the damage centers representing the full range of the flow-frequency-damage relationship for base conditions and for regulated conditions. Also, develop synthetic events that have consistency in volumes of runoff and peak flows and are reasonably representative regarding upstream contributions to downstream flows.

f. Perform sequential analysis with the developed hydrology. The period-of-record simulation provides simulated regulated flow which can be analyzed directly to develop flow-frequency relations. The representative flood approach requires an assumption that the regulated-flow frequency is the same as the natural-flow frequency. Frequency curves of regulated conditions at each damage center can then be derived from frequency curves of unregulated flows simply by assuming that a given ratio of the base flood will have the same recurrence frequency whether it is modified by regulatory structures or not. This assumption is valid as long as larger unregulated floods always correspond to the larger regulated flows.

g. Derive a flow-frequency and stage-discharge curve for the index station at each damage center as described in Chapters 6 and 8, for unregulated conditions for each plan of improvement. These can be used for determining average annual damage for unregulated conditions and for each plan of development and would thus form the primary basis for project selection.

h. Develop a PMF for each reservoir site, using procedures described in Chapter 7. These will be used as a possible basis for spillway design. Route the PMF through each reservoir, assuming reasonably adverse conditions for initial storage and available outlet capacity.

Chapter 11
Conservation Storage

11-1. General Considerations

a. Purposes. Water stored in the conservation pool can serve many purposes. The primary purposes for conservation storage are water supply, navigation, low-flow augmentation, fish and wildlife, and hydroelectric power. The water requirements for these purposes are discussed in this chapter along with water quality considerations. Methods for estimating the conservation storage, or yield, are presented in Chapter 12.

b. Operational policy. In general, the operational policy is to conserve available supplies and to release only when supplemental flow is needed to meet downstream requirements. Water stored in the conservation pool also provides benefits within the pool, such as lake recreation and fish and wildlife habitat.

c. Changing hydrology. When a reservoir is filled, the hydrology of the inundated area and its immediate surroundings is changed in a number of respects. The effects of inflows at the perimeter of the reservoir are translated rapidly to the reservoir outlet, thus, effectively speeding the flow of water through the reservoir. Also, large amounts of energy are stored and must be dissipated or utilized at the outlet. The reservoir loses water by evaporation, and this usually exceeds preproject evapotranspiration losses from the lake area. Siltation usually seals the reservoir bottom, but rising and falling water levels may alter the pattern of groundwater storage due to movement into and out of the surrounding reservoir banks. At high stages, water may seep from the reservoir through permeable soils into neighboring catchment areas and so be lost to the area of origin. Finally, sedimentation takes place in the reservoir and scour occurs downstream.

d. Storage allocation. The joint use of storage for more than one purpose creates problems of storage allocation for the various purposes. While retained in reservoir storage, water may provide benefits to recreation, fish, wildlife, hydropower, and aesthetics. Properly discharged from the reservoir, similar benefits are achieved downstream. Other benefits that can be derived from the reservoir are those covered in this chapter, including municipal and industrial water supply, agricultural water supply, navigation, and low-flow augmentation.

e. Supplemental storage capacity. In most areas, supplemental storage capacity is required for sediment deposition; otherwise, the yield capability of the reservoir may be seriously diminished during the project's economic life. Sediment storage is determined by estimating the average annual sediment yield per square mile of drainage area from observations in the region and multiplying by the drainage area and the economic life of the project. Trap efficiency of the reservoir is evaluated and the distribution of this estimated volume of sediment is determined, using methods described in EM 1110-2-4000 *Sedimentation Investigations of Rivers and Reservoirs*. Sediment surveys within the reservoir during actual operation will establish the reliability of these estimates. Storage allocation levels may then be revised if the sediment surveys show a significant difference between what was projected and what was measured. More complete descriptions of the techniques used to determine reservoir sedimentation are presented in EM 1110-2-4000.

f. Minimum pool. A minimum pool at the bottom of active conservation storage is usually established to identify the lower limit of normal reservoir drawdown. The inactive storage below the minimum pool level can be used for recreation, fish and wildlife, hydropower head, sediment deposition reserve, and other purposes. In rare instances, it might be used to relieve water supply emergencies.

g. Reservoir outlets. Reservoir outlets must be located low enough to withdraw water at desired rates with the reservoir stage at minimum pool. These outlets can discharge directly into an aqueduct or into the river. In the latter case, a diversion dam may be required downstream at the main canal intake.

h. Computing storage capacity. Because the primary function of reservoirs is to provide storage, their most important physical characteristic is storage capacity. Capacities of reservoirs on natural sites must usually be determined from topographic surveys. The storage capacity can be computed by planimetering the area enclosed within each elevation contour throughout the full range of elevations within the reservoir site. The increment of storage between any two elevation contours is usually computed by multiplying the average of the areas at the two elevations by the elevation difference. The summation of these increments below any elevation is the storage volume below that level. An alternative to the average-end-area method is the determination of the storage capacity by the conic method, which assumes that the volumes are more nearly represented by portions of a cone. This method is available in the HEC-1 *Flood Hydrograph Package* computer program and is described in the program user's manual. In the absence of adequate topographic maps, cross sections of the reservoir area are sometimes surveyed, and the capacity is computed from these vertical cross sections by using the formula for the volume of a prism.

11-2. Water Supply

a. Introduction. Water supply for any purpose is usually obtained from groundwater or from surface waters. Groundwater yields and the methods currently in use are covered in *Physical and Chemical Hydrogeology* (Domenico and Schwartz 1990). This discussion is limited to surface water supplies for low-flow regulation or for diversion to demand areas.

(1) In some cases, water supply from surface waters involves only the withdrawal of water as needed from a nearby stream. However, this source can be unreliable because streamflows can be highly variable, and the desired amount might not always be available. An essential requirement of most water supply projects is that the supply be available on a dependable basis. Reservoirs play a major role in fulfilling this requirement. Whatever the ultimate use of water, the main function of a reservoir is to stabilize the flow of water, either by regulating a varying supply in a natural stream or by satisfying a varying demand by the ultimate consumer. Usually, some overall loss of water occurs in this process.

(2) In determining the location of a proposed reservoir to satisfy water needs, a number of factors should be considered. The dam should be located so that adequate capacity can be obtained, social and environmental effects of the project will be satisfactory, sediment deposition in the reservoir and scour below the dam will be tolerable, the quality of water in the reservoir will be commensurate with the ultimate use, and the cost of storing and transporting the water to the desired location is acceptable. It is virtually impossible to locate a reservoir site having com-pletely ideal characteristics, and many of these factors will be competitive. However, these factors can be used as general guidelines for evaluating prospective reservoir sites.

(3) In the planning and design of reservoirs for water supply, the basic hydrologic problem is to determine how much water a specified reservoir capacity will yield. Yield is the amount of water that can be supplied from the reservoir to a specified location and in a specified time pattern. Firm yield is usually defined as the maximum quantity of water that can be guaranteed with some specified degree of confidence during a specific critical period. The critical period is that period in a sequential record that requires the largest volume from storage to provide a specified yield. Chapter 12 describes procedures for yield determination.

b. Municipal and industrial water use. The water requirement of a modern city is so great that a community system capable of supplying a sufficient quantity of potable water is a necessity. The first step in the design of a waterworks system is a determination of the quantity of water that will be required, with provision for the estimated requirements of the future. Next, a reliable source of water must be located and, finally, a distribution system must be provided. Water at the source may not be potable, so water-purification facilities are ordinarily included as an integral part of the system. Water use varies from city to city, depending on the population, climatic conditions, industrialization, and other factors. In a given city, use varies from season to season and from hour to hour. Planning of a water supply system requires that the probable water use and its variations be estimated as accurately as possible.

(1) Municipal uses of water may be divided into various classes. Domestic use is water used in homes, apartment houses, etc., for drinking, bathing, lawn and garden sprinkling, and sanitary purposes. Commercial and industrial use is water used by commercial establishments and industries. Public use is water required in parks, civic buildings, schools, hospitals, churches, street washing, etc. Water that leaks from the system, unauthorized connections, and other unaccounted-for water is classified as loss and waste.

(2) The average daily use of water for municipal and industrial purposes is influenced by many factors. More water is used in warm, dry climates than in humid climates for bathing, lawn watering, air conditioning, etc. In extremely cold climates water may be wasted at faucets to prevent freezing of pipes. Water use is also influenced by the economic status of the users. The per capita use of water in slum areas is much less than that in high-cost residential districts. Manufacturing plants often require large amounts of water; however, some industries develop their own water supply and place little or no demand on a municipal system. The actual amount depends on the extent of the manufacturing and the type of industry. Zoning of the city affects the location of industries and may help in estimating future industrial demands.

(3) About 80 percent of industrial water may be used for cooling and need not be of high quality, but water used for process purposes must be of good quality. In some cases, industrial water must have a lower content of dissolved salts than can be permitted in drinking water.

The location of industry is often much influenced by the availability of water supply. If water costs are high, less water is used, and industries will often develop their own supply to obtain cheaper water. In this respect, the installation of water meters in some communities has reduced water use by as much as 40 percent. The size of the city being served is a factor affecting water use. Per capita use tends to be higher in large cities than in small towns. The difference results from greater industrial use, more parks, greater commercial use, and, perhaps, more loss and waste in the larger cities. All of these factors, plus estimated population projections, should be considered in designing a waterworks system.

(4) The use of water in a community varies almost continually. In midwinter the average daily use is usually about 20 percent lower than the daily average for the year, while in summer it may be 20 to 30 percent above the daily average for the year. Seasonal industries such as canneries may cause wide variation in water demand during the year. It has been observed that for most communities, the maximum daily use will be about 180 percent of the average daily use throughout the year. Within any day, large variations can be as low as 25 percent to as high as 200 percent of the average for portions of the day. The daily and hourly variations in water use are not usually considered in reservoir design, because most communities use distribution reservoirs (standpipes, etc.) to regulate for these variations.

c. *Agricultural water use.* The need for agricultural water supply is primarily for irrigation. Irrigation can be defined as the application of water to soil to supplement deficient rainfall in order to provide moisture for plant growth. In the United States, about 46 percent of all the water used is for irrigation. Irrigation is a consumptive use; that is, most of the water is transpired or evaporated and is essentially lost to further use.

(1) In planning an irrigation project a number of factors must be considered. The first step would be to establish the capability of the land to produce crops that provide adequate returns on the investment in irrigation works. This involves determining whether the land is arable (land which, when properly prepared for agriculture, will have a sufficient yield to justify its development) and irrigable.

(2) The amount of water required to raise a crop depends on the kind of crop and the climate. The plants that are the most important sources of food and fiber need relatively large amounts of water. The most important climatic characteristic governing water need is the length of the growing season. Other factors that affect water requirements are the quality of the water, the amount of land to be irrigated, and, of course, the cost of the water to the irrigator.

(3) In estimating the amount of storage that will be required in a reservoir for irrigation, the losses and waste that occur in the irrigation system must be considered. Losses and waste are usually divided into conveyance and irrigation losses and waste. Conveyance losses and waste are those that occur in the conveyance and distribution system prior to the application of water to crops. These are dependent on the design and construction of the system and also on how the system is operated and maintained. Irrigation losses and waste are those that occur due to the slope of the irrigated land, the preparation of the land, soil condition, the method of irrigation, and the practices of the irrigator.

(4) Usually, most of the irrigation losses and waste, as well as a portion of the water applied to the irrigated lands, return to the stream. If there are requirements for flow downstream of the reservoir, these return flows can be important in determining the amount of water that must be released to meet such requirements.

(5) In most areas, the need for irrigation water is seasonal and depends on the growing season, the number of crops per year, and the amount of precipitation. For these reasons the variation of the demand is often high, ranging from no water for some months up to 20 to 30 percent of the annual total for other months. This variation can have a very large effect on the amount of storage required and the time of year when it is available.

11-3. Navigation and Low-Flow Augmentation

a. *Objective.* In designing a reservoir to supply water for navigation and low-flow augmentation, the objective is significantly different from objectives for the other purposes that have been discussed previously in this chapter. The objective is to supplement flows at one or more points downstream from the reservoir. For navigation, these flows aid in maintaining the necessary depth of water and alleviate silting problems in the navigable channel. Low-flow augmentation serves a number of purposes including recreation, fish and wildlife, ice control, pollution abatement, and run-of-river power projects. Under certain conditions, low-flow augmentation provides water for the other purposes discussed in this chapter. For instance, if the intake for a municipal and industrial water supply is at some point downstream of the reservoir, the objective may be to supplement low flows at that point.

b. Criteria for navigability. There are no absolute criteria for navigability and, in the final analysis, economic criteria control. The physical factors that affect the cost of waterborne transport are depth of channel, width and alignment of channel, locking time, current velocity, and terminal facilities. Commercial inland water transport is, for the most part, accomplished by barge tows consisting of 1 to 10 barges pushed by a shallow-draft tug. The cost of a trip between any two terminals is the sum of the fuel costs and wages, fixed charges, and other operating expenses depending on the time of transit. Reservoirs aid in reducing these costs by providing the proper depth of water in the navigation channel, or by providing a slack-water pool in lock and dam projects. Storage reservoirs can rarely be justified economically for navigation purposes alone and are usually planned as multipurpose projects. Improving navigation by using reservoirs is possible when flood flows can be stored for release during low-flow seasons.

c. Supplying deficiencies without waste. The ideal reservoir operation for navigation or low-flow augmentation would provide releases so timed as to supply the deficiencies in natural flow without waste. This is possible only if the reservoir is at the head of a relatively short control reach. As the distance from the reservoir to the reach is increased, releases must be increased to allow for uncertainties in estimating intermediate runoff and for evaporation and seepage enroute to the reach to be served. Moreover, the releases must be made sufficiently far in advance of the need to allow for travel time to the reach, and in sufficient quantity so that after reduction by channel storage, the delivered flows are adequate. The water requirement for these releases is considerably greater than the difference between actual and required flows.

d. Climate. Climate can also affect reservoir operation for low-flow regulation. Depending on the purpose to be served, the releases may be required only at certain times of the year or may vary from month to month. For pollution abatement, the important factors are the quality of the water to be supplemented, the quality of the water in the reservoir, and the quality standard to be attained. Also, the level of the intakes from which releases will be made can be a very sufficient factor in pollution abatement, since the quality can vary from one level to another in the reservoir. Long-term variations can occur due to increased contamination downstream of a reservoir. This should be considered in determining the required storage in the reservoir.

11-4. Fish and Wildlife

a. Added authorized purposes. As shown in Figure 2-1, fish and wildlife and subsequent environmental purposes have been added as authorized purposes since 1960. Because many of the reservoirs were built prior to that time, their authorized purposes and regulation plans may not adequately reflect the more recent environmental objectives. Therefore, there is an increasing demand and need for the evaluation of environmental impacts for these projects.

b. Water level fluctuations. The seasonal fluctuation that occurs at many flood control reservoirs and the daily fluctuations that occur with hydropower operation often result in the elimination of shoreline vegetation and subsequent shoreline erosion, water quality degradation, and loss of habitat for fish and wildlife. Adverse impacts of water level fluctuations also include loss of shoreline shelter and physical disruption of spawning and nests.

c. Water level management. Water-level management in fluctuating warm-water and cool-water reservoirs generally involves raising water levels during the spring to enhance spawning and the survival of young predators. Pool levels are lowered during the summer to permit regrowth of vegetation in the fluctuation zone. Fluctuations may be timed to benefit one or more target species; therefore, several variations in operation may be desirable. In the central United States, managers frequently recommend small increases in pool levels during the autumn for waterfowl management.

d. Fishery management. Guidelines to meet downstream fishery management potentials are developed based on project water quality characteristics and water control capabilities. To do so, an understanding of the reservoir water quality regimes is critical for developing the water control criteria to meet the objectives. For example, temperature is often one of the major constraints of fishery management in the downstream reach, and water control managers must understand the temperature regime in the pool and downstream temperature requirements, as well as the capability of the project to achieve the balance required between the inflows and the releases. Releasing cold water downstream where fishery management objectives require warm water will be detrimental to the downstream fishery. Conversely, releasing warm water creates difficulty in maintaining a cold-water fishery downstream.

e. Water temperature management. Water control activities can also impact water temperatures within the pool by changing the volume of water available for a particular layer. In some instances, cold-water reserves may be necessary to maintain a downstream temperature objective in the late summer months; therefore, the availability of cold water must be maintained to meet this objective. For some projects, particularly in the southern United States, water control objectives include the maintenance of warm-water fisheries in the tailwaters. In other instances, fishery management objectives may include the maintenance of a two-story fishery in a reservoir, with a warm-water fishery in the surface water, and a cold-water fishery in the bottom waters. Such an objective challenges water control managers to regulate the project to maintain the desired temperature stratification while maintaining sufficient dissolved oxygen in the bottom waters for the cold-water fishery. Regulation to meet this objective requires an understanding of operational affects on seasonal patterns of thermal stratification, and the ability to anticipate thermal characteristics.

f. Minimum releases. Minimum instantaneous flows can be beneficial for maintaining gravel beds downstream for species that require this habitat. However, dramatic changes in release volumes, such as those that result from flood-control regulation, as well as hydropower, can be detrimental to downstream fisheries. Peaking hydropower operations can result in releases from near zero to very high magnitudes during operations at full capacity. Maintaining minimum releases and incorporating reregulation structures are two of the options available to mitigate this problem.

g. Fishing versus peak power. In some instances, tailwater fishing is at a maximum during summer weekends and holidays, and this is a time when power generation may be at a minimum and release near zero. Maintaining minimum releases during weekend daylight hours may improve recreational fishing, but may reduce the capability to meet peak power loads during the week because of lower water level (head) in the reservoir. In these instances, water control managers will be challenged to regulate the project with consideration of these two objectives.

h. Anadromas fish. Regulation for anadromous fish is particularly important during certain periods of the year. Generally, upstream migration of adult anadromous fish begins in the spring of each year and continues through early fall, and downstream migration of juvenile fish occurs predominantly during the spring and summer months. The reduced water velocities through reservoirs, in comparison with preproject conditions, may greatly lengthen the travel time for juvenile fish downstream through the impounded reach. In addition, storage for hydropower reduces the quantity of spill, and as a result, juvenile fish must pass through the turbines. The delay in travel time subjects the juvenile fish to greater exposure to birds and predator fish, and passage through the powerhouse turbines increases mortality. To improve juvenile survival, storage has been made available at some projects to augment river flows, and flows are diverted away from the turbine intakes and through tailraces where the fish are collected for transportation or released back into the river. Barges or tank trucks can be used to transport juveniles from the collector dams to release sites below the projects. Other Corps projects have been modified so the ice and trash spillways can be operated to provide juvenile fish passage.

i. Wildlife habitat. Project regulation can influence wildlife habitat and management principally through water level fluctuations. The beneficial aspects of periodic drawdowns on wildlife habitat are well documented in wildlife literature. Drawdowns as a wildlife management technique can, as examples, allow the natural and artificial revegetation of shallows for waterfowl, permit the installation and maintenance of artificial nesting structures, allow the control of vegetation species composition, and ensure mast tree survival in greentree reservoirs. Wildlife benefits of periodic flooding include inhibiting the growth of undesirable and perennial plants, creating access and foraging opportunities for waterfowl in areas such as greentree reservoirs, and ensuring certain water levels in stands of vegetation to encourage waterfowl nesting and reproduction.

11-5. Hydroelectric Power

a. General. The feasibility of hydroelectric development is dependent upon the need for electric power, the availability of a transmission system to take the power from the point of generation to the points of demand, and the availability of water from streamflow and storage to produce power in accordance with the capacity and energy demands in the power market area. Also, the project's power operations must be coordinated with the operations for other project purposes to ensure that all purposes are properly served. Each of these factors must be investigated to ensure that the project is both feasible and desirable and to minimize the possibility that unforeseen conflicts will develop between power and other water uses during the project life.

(1) The ability of a project to supply power is measured in terms of two parameters: **capacity** and **energy**. **Capacity**, commonly measured in kilowatts (kw), is the maximum amount of power that a generating plant can deliver. **Energy**, measured in kilowatt-hours (kwh), is the

amount of actual work done. Both parameters are important, and the operation of a hydroelectric project is sensitive to changes in the demand for either capacity or energy.

(2) Experience has indicated that it is very unlikely that power demands will remain unchanged during the project life. Furthermore, the relative priority of various other water uses can change during the project life, and there are often legal, institutional, social, or environmental factors that might affect the future use of water at a particular project. Consequently, the feasibility studies for a proposed project must not be limited to conditions that are only representative of the current time or the relatively near future. Instead, the studies must include considerations of future conditions that might create irreconcilable conflicts unless appropriate remedial measures are provided for during project formulation.

(3) This section presents general concepts for the hydrologic analyses associated with the planning, design, and operation of hydroelectric projects and systems. More detailed information is provided in EM 1110-2-1701. Other investigations that influence or affect the hydrologic studies will be discussed to the extent that their outcome must be understood by the hydrologic engineer.

b. Types of hydroelectric load. Power developments, for purposes of this discussion, can be classified with respect to the type of load served or the type of site development proposed. The two categories related to the type of load served are baseload and peaking plants.

(1) Base load. Baseload plants are projects that generate hydroelectric power to meet the baseload demand. The baseload demand is the demand that exists 100 percent of the time. The baseload can readily be seen in Figure 11-1 as the horizontal dashed line on a typical annual load duration curve. This curve displays the percent of time during a given year that a given capacity demand is equaled or exceeded. The area under this curve represents the total energy required to meet the load during the year. Usually, the baseload demand is met by thermal generating facilities. However, in cases where there is a relatively abundant supply of water that is available with a high degree of reliability and where fuel is relatively scarce, hydroelectric projects may be developed to meet the baseload demands. These projects would then operate at or near full capacity 24 hr per day for long periods of time. This type of development is not feasible where there is a large seasonal variation in streamflow unless the baseflow is relatively high or unless there is a provision for a large volume of power storage in the project.

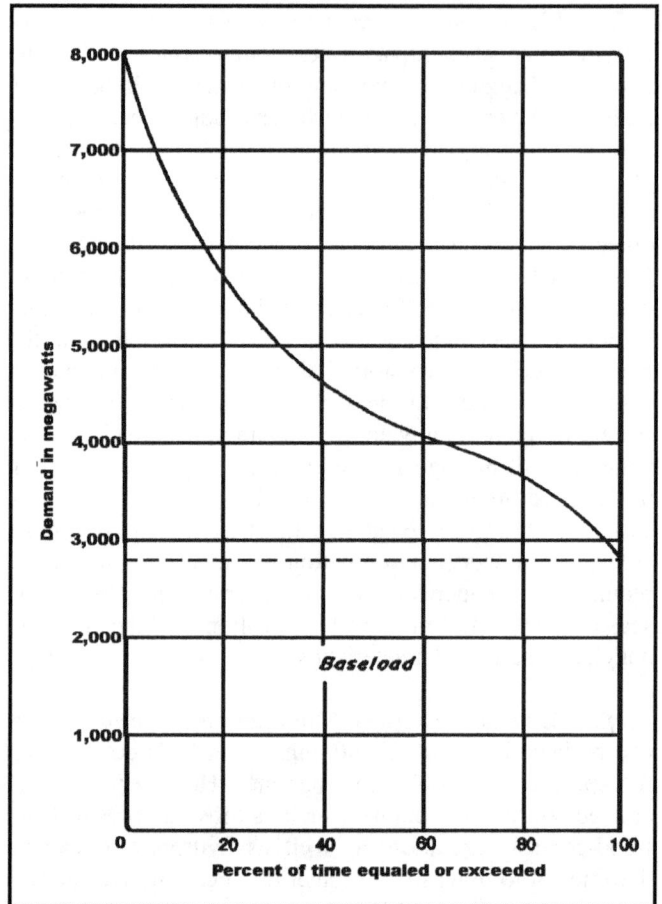

Figure 11-1. Typical annual load duration curve

(2) Peaking load. Peaking plants are projects that generate hydroelectric power to supplement baseload generation during periods of peak power demands. The peak power demands are the loads that exist primarily during the daylight hours. The time of occurrence and magnitude of peak power demands are shown on a load curve in Figure 11-2. This curve shows the time variation in power demands for a typical week. Depending upon the quantity of water available and the demand, a peaking plant may generate from as much as 18 hr a day to as little as no generation at all, but it is usually 8 hr a day or less. Peaking plants must supply sufficient capacity to satisfy the peak capacity demands of a system and sufficient energy to make the capacity usable on the load. This means that energy or water should be sufficient to supply peaking support for as long and as often as the capacity is needed. In general, a peaking hydroelectric plant is desirable in a system that has thermal generation facilities to meet the baseload demands. The hydroelectric generating facilities are particularly adaptable to the peaking operation because

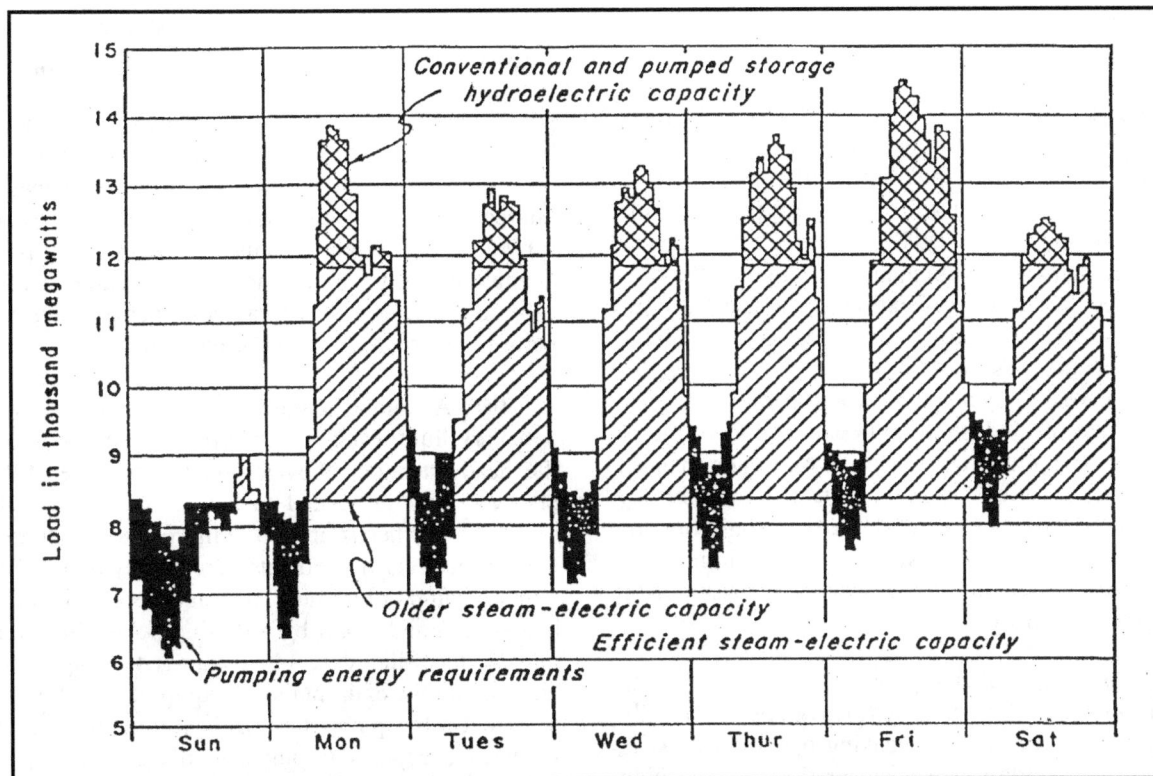

Figure 11-2. Weekly load curve for a large electric system

their loading can be changed rapidly. Also, the factors that make seasonal variations in streamflow a major problem in baseload operation are usually quite easily overcome in a peaking plant if some storage can be provided.

c. Project types. There are three major classifications of hydroelectric projects: storage, run-of-river, and pumped storage. There are also combinations of projects that might be considered as separate classifications, but for purposes of discussing hydrologic analysis it is necessary to define only these three types.

(1) Storage projects. Storage plants are projects that usually have heads in the medium to high range (> 25 m) and have provisions for storing relatively large volumes of water during periods of high streamflow in order to provide water for power generation during periods of deficient streamflow. Considerable storage capacity may be required because the period of deficient flow is quite frequently more than a year long and, in some instances, may be several years long. Because use of the stored water entails drawdown of the power storage, it is desirable that other water uses associated with the development of a storage plant permit frequent and severe drawdowns during dry periods. Peaking operation, which is quite frequently associated with storage projects requires large

and sometimes rapid fluctuations in releases of water through the generating units. It is often necessary to provide facilities to re-regulate the power releases if fluctuations of water levels below the project are not tolerable. Because storage projects are conducive to multipurpose use and because the power output from a storage plant is a function of the guaranteed output during a multi-year dry period, it is usually necessary to make detailed routing studies to determine the storage requirements, installed capacity, firm energy, and an operating plan.

(2) Run-of-river projects. Run-of-river plants have little or no power storage and, therefore, must generate power from streamflow as it occurs. The projects generally have productive heads in the low to medium range (5-30 m) and are quite frequently associated with navigational developments or other multipurpose developments with limitations on reservoir drawdowns. Because of the absence or near-absence of storage in run-of-river projects, there is usually very little operational flexibility in these projects, and it is necessary that all water uses be compatible. The existence of one or more storage projects in the upstream portion of a river basin may make a run-of-river project in the lower portion of the basin feasible where it would not otherwise be feasible. In this situation, the storage projects provide a regulated outflow

11-7

that is predictable and usable, while the natural streamflow might be neither.

(a) Run-of-river projects may have provisions for a small amount of storage, often called pondage. This pondage detains the streamflow during off-peak periods in daily or weekly cycles for use in generating power during peak demand periods. If the cycle of peaking operation is a single day, the pondage requirements are based on the flow volume needed to sustain generation at or near installed capacity for 12 hr. If more storage capacity is available and large fluctuations in the reservoir surface are permissible, a weekly cycle of peaking operation may be considered. Because industrial and commercial consumption of power is significantly lower on weekends than on week days, an "off-peak" period is created from Friday evening until Monday morning. If generation from the hydroelectric peaking plants is not required during this period, water can be stored in the pondage for use during the 5-day peak-load period.

(b) Because of the relatively low heads associated with run-of-river projects, the tailwater fluctuations are usually quite important, particularly in peaking operations. Also, flood flows may curtail power generation due to high tailwater. While flow-duration analysis can be used to estimate average annual energy production, sequential analysis may be required for more detailed analysis of extreme conditions.

(3) Pumped-storage projects. Pumped-storage plants are projects that depend on pumped water as a partial or total source of water for generating electric energy. This type of project derives its usefulness from the fact that the demand for power is generally low at night and on weekends; therefore, pumping energy at a very low cost will be available from idle thermal generating facilities or run-of-river projects. If there is a need for peaking capacity and if the value of peaking power generation sufficiently exceeds the cost of pumping energy (at least a ratio of 1.5 to 1.0 because roughly 3 kwh of pumping energy are necessary to deliver enough water to provide 2 kwh of energy generation), pumped storage might be feasible. There are three types of pumped-storage development: diversion, off-channel, and in-channel, which are detailed in Chapter 7 of EM 1110-2-1701.

(a) In general, pumped storage projects consist of a high-level forebay where pumped water is stored until it is needed for generation and a low-level afterbay where the power releases are regulated, if necessary, and from which the water is pumped. The pumping and generating are done by generating units composed of reversible pump turbines and generator motors located along a tunnel or penstock connecting the forebay and afterbay. The water is pumped from the afterbay to the forebay when the normal power demand is low and least expensive and released from the forebay to the afterbay to generate power when the demand is high and most costly. The feasibility of pumped-storage developments is dependent upon the need for relatively large amounts of peaking capacity, the availability of pumping energy at a guaranteed favorable cost, and a load with an off-peak period long enough to permit the required amount of pumping.

(b) A unique feature of pumped-storage systems is that very little water is required for their operation. Once the headwater and tailwater pools have been filled, only enough water is needed to take care of evaporation and seepage. For heads up to 300 m, reversible pump turbines have been devised to operate at relatively high efficiency as either a pump or turbine. The same electrical unit serves as a generator and motor by reversing poles. Such a machine may reduce the cost of a pumped-storage project by eliminating the extra pumping equipment and pump house. The reversible pump turbine is a compromise in design between a Francis turbine and a centrifugal pump. Its function is reversed by changing the direction of rotation.

d. Need for hydroelectric power. The need for power is established by a power market study or survey. The feasibility of a particular hydroelectric project or system is determined by considering the needs as established by the survey, availability of transmission facilities, and the economics of the proposed project or projects. Although forecasts of potential power requirements within a region to be served by a project are not hydrologic determinations, they are essential to the development of plans for power facilities and to the determination of project feasibility and justification. The power market survey is a means of evaluating the present and potential market for electrical power in a region.

(1) The survey must provide a realistic estimate of the power requirements to be met by the project and must show the anticipated rate of load growth from initial operation of the project to the end of its economic life. The survey also provides information regarding the characteristics of the anticipated demands for power. These characteristics, which must be considered in hydrologic evaluations of hydroelectric potential, include the seasonal variation of energy requirements (preferably on a monthly basis), the seasonal variation of capacity requirements (also preferably on a monthly basis), and the range of usable plant factors for hydroelectric projects under both adverse and average or normal flow conditions.

(2) The results of a power market survey might be furnished to the hydrologic engineer in the form of load duration or load curves (Figures 11-1 and 11-2) showing the projected load growth, the portion of the load that can be supplied by existing generating facilities, and the portion that must be supplied by future additions to the generating system. From these curves, the characteristics of planned hydroelectric generating facilities can be determined. Because these data are developed from the needs alone without consideration of the potential for supplying these needs, the next step is to study the potential for hydroelectric development, given the constraints established in the study of needs.

e. Estimation of hydroelectric power potential. Traditionally, hydroelectric power potential has been determined on the basis of the critical hydro-period as indicated by the historical record. The critical hydro-period is defined as the period when the limitations of hydroelectric power supply due to hydrologic conditions are most critical with respect to power demands. Thus, the critical period is a function of the power demand, the streamflow, and the available storage. In preliminary project planning, the estimates of power potential are often based on a number of simplifying assumptions because of the lack of specific information for use in more detailed analyses. Although these estimates and the assumptions upon which they are based are satisfactory for preliminary investigations, they are not suitable for every level of engineering work. Many factors affecting the design and operation of a project are ignored in these computations. Therefore, detailed sequential analyses of at least the critical hydro-period should be initiated as early as possible, usually when detailed hydrologic data and some approximate physical data concerning the proposed project become available. Because of the availability of computer programs for accomplishing these sequential routings, they can be done rapidly and at a relatively low cost.

(1) The manner in which the streamflow at a given site is used to generate power depends upon the storage available at the site, the hydraulic and electrical capacities of the plant, streamflow requirements downstream from the plant, and characteristics of the load to be served. In theory, the hydroelectric power potential at a particular site, based on repetition of historical runoff, can be estimated by identifying the critical hydro-period and obtaining estimates of the average head and average streamflow during this critical period. The data can then be used in the equation below to calculate the power available from the project:

$$kW = \frac{1}{11.81} QHe \qquad (11\text{-}1)$$

In order to convert a project's power output to energy, Equation 11-1 must be integrated over time:

$$kWh = \frac{1}{11.81} \int_0^t Q_t H_t e\, dt \qquad (11\text{-}2)$$

where

kW = power available from the project, kW

kWh = energy generated during a time period, kWh

Q = average streamflow during the time period, m^3/sec

H = average head during the time period, m (Head = headwater elevation - tailwater elevation - hydraulic losses)

t = number of hours in the time period

e = overall efficiency expressed as the product of the generator efficiency and the turbine efficiency

In practice, the summation of energy production over the critical period is performed with a sufficiently small time step to provide reasonable estimates of head and, therefore, energy. Two basic approaches are available: flow-duration and sequential analysis.

(2) For run-of-river projects, where the headwater elevation does not vary significantly, the flow-duration approach can be used to estimate average annual energy production. The duration curve can be truncated at the minimum flow rate for power production. The curve can also be truncated for high-flows if the tailwater elevation is too high for generation. The remaining curve is converted to capacity-duration and integrated to obtain average annual energy. Hydropower Analysis Using Flow-Duration Procedures HYDUR (HEC 1982d) was developed to perform energy computations based on flow-duration data. EM 1110-2-1701 describes HYDUR in paragraph C-2*b* and the flow-duration method in Section 5-7.

(3) Sequential streamflow analysis will be applied to most reservoir studies. The procedure allows detailed computations of the major parameters affecting hydro-power (e.g., headwater and tailwater elevation, efficiency, and flow release). By performing the analysis in sufficiently small time steps, an accurate simulation of the reservoir operation, power capacity and energy production can be obtained. Chapter 5 of EM 1110-2-1701, Sections 5-8 through 5-10, provides a discussion of sequential routing studies. Appendix C provides information concerning computer programs that are available for use in these studies.

f. Hydropower effect on other project purposes. Usually, power generation must have a high priority relative to other conservation uses. Consequently, thorough investigations of all aspects of the power operation must be conducted to ensure that the power operations do not create intolerable situations for other authorized or approved water uses. Likewise, the power operations must be coordinated with other high priority purposes such as flood control and municipal water supply to ensure that the planned power operation will not interfere with the operations for these purposes. The operation rules that are necessary to effect the coordination are usually developed and tested using engineering judgment and detailed sequential routing studies. However, it is necessary to define the interactions between power and other project purposes before initiating operation studies.

(1) Power generation is generally compatible with most purposes that require releases of water from a reservoir for downstream needs. However, power generation usually competes with purposes that require withdrawal of the water directly from the reservoir or that restrict fluctuations in the reservoir level. Flood-control requirements frequently conflict with power operations because flood-control needs may dictate that storage space in a reservoir be evacuated at a time when it would be beneficial to store water for use in meeting future power demands. Furthermore, when extensive flooding is anticipated downstream from a reservoir project, it may be necessary to curtail power releases to accomplish flood-control objectives. It is often possible to pass part or all of the flood-control releases through the generating units, thereby reducing the number of additional outlets needed and significantly increasing the energy production over what would be possible if the flood-control releases were made through conduits or over the spillway. Also, many of the smaller floods can be completely regulated within the power drawdown storage, an operation that is beneficial to power because it provides water for power generation that might otherwise have been spilled. This joint use can reduce the exclusive flood-control storage

requirements and also reduce the frequency of use of flood-control facilities.

(2) Water for municipal, industrial, or agricultural use can be passed through the generating units with no harmful effects if the point of withdrawal for the other use is below the point where the power discharge enters the river. Re-regulation may be required for hydropower peaking operations to "smooth out" the power releases. Conflicts between power and these consumptive uses more likely occur when the withdrawal for other uses is directly from the reservoir. When the withdrawal is from the reservoir of a storage project, the inclusion of power as a project purpose may require that special attention be given to intake facilities for the other purposes because of the relatively large drawdown associated with storage projects.

(3) Low-flow augmentation for navigation, recreation, or fish and wildlife can be accomplished by releases through power generating units. In the case of baseload projects, the power release is ideally suited for this type of use. With peaking projects, however, a re-regulation structure may be necessary to provide the relatively uniform releases that might be required for navigation or for in-stream recreation. Release of water for quality enhancement can sometimes be accomplished through the generating units. Although the intakes for the turbines are usually located at a relatively low elevation in the reservoir where dissolved oxygen content might be low, the oxygenation that occurs in the tailrace and in the stream below the project may produce water with an acceptable dissolved oxygen content. The water released from the lower levels of the reservoir is normally at a relatively low temperature and, thus, ideal for support or enhancement of a cold-water fishery downstream. If warm waters are needed for in-stream recreation, for fishery requirements, or for any other purpose, a special multilevel intake may be required to obtain water of the desired temperature.

(4) Recreation values at a reservoir project may be enhanced, somewhat, by the inclusion of power because a much larger reservoir is frequently required, and that may increase opportunities for extensive recreational activities. Unfortunately, however, the large drawdowns associated with the big storage projects create special problems with respect to the location of permanent recreational facilities and may create mudflats that are undesirable from the standpoint of aesthetics and public health requirements. The drawdown may also expose boaters, swimmers, and other users to hazardous underwater obstacles unless provisions are made to remove these obstacles to a point well below the maximum anticipated drawdown. Obviously the time of occurrence of extreme drawdown

conditions is an important factor in determining the degree of conflict with recreation activities.

11-6. Water Quality Considerations

a. Water quality definition. Water quality deals with the kinds and amounts of matter dissolved and suspended in natural water, the physical characteristics of the water, and the ecological relationships between aquatic organisms and their environment. It is a term used to describe the chemical, physical, and biological characteristics of water in respect to its suitability for a particular purpose. The same water may be of good quality for one purpose or use, and bad for another, depending on its characteristics and the requirements for the particular use.

b. Water quality parameters. In general, physical parameters define the water quality characteristics that affect our senses while chemical and biological parameters index the chemical and biological constituents present in the water resource system. However, these are not independent, but are actually highly related. For example, chemical waste discharges may affect such physical factors as density and color, may alter chemical parameters such an pH and alkalinity, and may affect the biological community in the water. Even so, the physical, chemical, and biological subdivision is a useful way to discuss water quality conditions. The following sections describe the water quality parameters frequently associated with reservoirs. EM 1110-2-1201 provides details on parameters, assessment techniques, plus data collection and analysis.

11-7. Water Quality Requirements

A wide variety of demands are made for the use of water resources. Water of a quality that is unsatisfactory for one use may be perfectly acceptable for another. The level of acceptable quality is often governed by the scarcity of the resource or the availability of water of better quality.

a. Domestic use. The use of water for domestic purposes such as drinking, culinary use, and bathing is generally considered to be the most essential use of our water resources. The regulations for the quality of this water are likewise higher than for most (but not all) other beneficial uses of water. In early times, the quality of the water supply source and the quality at the delivery point were synonymous; but the general degradation of both surface water quality and shallow ground water quality has made it necessary, in most cases, for some degree of water treatment to be used to produce acceptable water for domestic use. In recent decades, there has been a strong trend (which is likely to continue) for the quality of the source waters throughout the world to decline as a result of increased urbanization and industrialization and as a result of changes in agricultural practices. At the same time, populations are coming to expect a higher standard of health and well-being; and as a result, the regulations for acceptable domestic water continue to rise and enlarge the role of water treatment.

b. Drinking water standards. Drinking water standards for the world as a whole have been set by the World Health Organization (WHO). One should keep in mind that these standards do not describe an ideal or necessarily desirable water, but are merely the maximum values of contaminant concentration which may be permitted. It is highly desirable to have water of much better quality. In the United States, the Environmental Protection Agency (EPA) sets regulations that legally apply to public drinking water and water supply systems. The regulations are divided into three categories: bacterial, physical, and chemical characteristics. They are defined in terms of maximum contaminant levels (MCL's). Bacterial quality is defined by establishing the sampling sequence, the method of analysis, and the interpretation of test results for the coliform organisms which serve as presumptive evidence of bacterial contamination from intestinal sources. Analysis is generally made for total coliform, fecal coliform, and streptococci coliform. The limits on biological and physical parameters, and on chemical elements or compounds in water are documented in *Water Supply and Sewerage* (McGhee 1991).

c. Quality of source waters. The drinking water standards are the end product of a production line which begins with the source water as a raw material and proceeds through the various unit processes of water treatment and finally water distribution. The quality of source waters for other uses such as agricultural and industrial water supply, fish and other aquatic life, and recreation are set by state regulations of receiving waters. Other specific uses may include regulations for navigation, wild and scenic rivers, and other state-specific needs. The state regulations are subject to EPA approval. The regulations of a given state may take a variety of forms but are often specified by stream reaches including associated natural or constructed impoundments. Each reach may be classified for its various water uses and water quality standards defined for each use.

(1) Industry uses water as a buoyant transporting medium, cleansing agent, coolant, and as a source of steam for heating and power production. Often the quality required for these purposes is significantly higher than that required for human consumption. The availability of water of high quality is often an important parameter in site

selection by an industry. The needs of a particular industry as to both quantity and quality of water varies with the competition for water, the efficiency of the plant process with regard to water utilization, the recycling of water, the location of the plant site and the ratio of the cost of the water to the cost of the product. For economic reasons and for reasons of quality control and operation responsibility, industries with high water requirements usually develop their own supply and treatment facilities.

(2) Farmstead water is that water used by the human farm population for drinking, food preparation, bathing, and laundry. It also includes water used for the washing and hydrocooling of fruits and vegetables, and water used in the production of milk. The quality of water desired for farmstead use is generally that required for public water supplies. It is not feasible to set rigid quality standards for irrigation waters because of such varied and complex factors as soil porosity, soil chemistry, climatic conditions, the ratio of rainwater to irrigation water, artificial and natural drainage, relative tolerance of different plants, and interferences between and among constituents in the water. Examples of the latter are the antagonist influence of calcium-sodium, boron-nitrates, and selenium-sulfates.

(3) Water quality parameters of importance for irrigation are sodium, alkalinity, acidity, chlorides, bicarbonates, pesticides, temperature, suspended solids, radionuclids, and biodegradable organics. All of these factors need to be weighed carefully in evaluating the suitability of water for irrigating a particular crop. There is surprisingly little data on the effect of water quality on livestock, but generally they thrive best on water meeting human drinking water standards. The intake of highly mineralized water by animals can cause physiological disturbances of varying degrees of severity. In some cases, particular ions such as nitrates, fluorides, selenium salts, and molybdenum may be harmful. Certain algae and protozoa have also been proven toxic to livestock.

(4) The basic purpose of water quality criteria for aquatic life is to restore or maintain environmental conditions that are essential to the survival, growth, reproduction, and general well-being of the important aquatic organisms. These criteria are ordinarily determined without the aid of economic considerations. Generally, a number of major problems arise in establishing water quality criteria for an aquatic community because of the inability to quantify the effects of the pertinent parameters and reduce them to a conceptual model that describes the nature of the biological community which will develop under a given set of conditions. But extensive research should be considered when unusually high concentrations of such parameters as alkalinity, acidity, heavy metals,

cyanides, oil, solids, turbidity, or insecticides are known to exist.

(5) For an aquatic system to be acceptable for swimming and bathing, it must be aesthetically enjoyable (i.e., free from obnoxious floating or suspended substances, objectionable color and foul odors), it must contain no substances that are toxic upon ingestion or irritating to the skin or sense organs, and it must be reasonably free of pathogenic bacteria. Standards generally do not cover the first two terms as related to swimming and bathing except in qualitative terms. In the United States, numerous standards exist for bacteriological quality based on the coliform count in the water. Generally, the standards range downward from 1,000 coliform bacteria per 100 ml to as low as 50 per 100 ml. Such standards are not based on demonstrated transmittal of disease, but have been established because in these ranges the standards are economically reasonable and no problems appear. The use of an aquatic system for boating and aesthetic enjoyment is generally not so demanding as the requirements for swimming or the propagation of fish and aquatic life although these three water uses are usually closely linked.

11-8. Reservoir Water Quality Management

Reservoirs may serve several purposes in the management of water quality. If used properly, substantial benefits can be achieved. On the other hand, unwise use of reservoirs may cause increased quality degradation. Benefits may accrue as a result of detention mixing or selective withdrawal of water in a reservoir or the blending of waters from several reservoirs. The effects of improper management are often far-reaching and long-term. They may range from minor to catastrophic, and may be as obvious as a fish kill or subtle and unnoticed. It is essential that all water control management activity and especially real-time actions include valid water quality evaluation as a part of the daily water control decision process. It must be understood that water quality benefits accumulate slowly, build on each other, and can become quite substantial over time. This is in contrast to the sudden benefits that come from a successful flood-control operation. Water quality management requirements, objectives, and standards are presented in EM 1110-2-3600.

a. *Reservoirs in streams.* The presence of a reservoir in a stream affects the quality of the outflow as compared to the inflow by virtue of the storage and mixing which takes place in the reservoir. The effect of such an impoundment may be easily evaluated for conservative parameters if the waters of the reservoir are sufficiently mixed that an assumption of complete mixing within an analysis time period does not lead to appreciable error. However, this assumption is limited to relatively small, shallow reservoirs.

b. Reservoir outflow and inflow. The simplest technique requires the assumption that the reservoir outflow during a given time period is of constant quality and equal to the quality of the reservoir storage at the end of the computation time period. It is then assumed that the inflow for the time period occurs independently of the outflow, and reservoir quality is determined by a quality mass balance at the end of the time period. This approach is equivalent to the mass balance of water in reservoir routing.

c. Reservoir water quality. Simple mass balance procedures may be applicable is some situations; however, usually more comprehensive methods should be considered. Chapter 4, "Water Quality Assessment Techniques," in EM 1110-2-1201 describes various techniques available for assessing reservoir water quality conditions. There is a hierarchy of available techniques that reflects increasing requirements of time, cost, and technical expertise. The increasing efforts should provide accompanying increases in the degree of understanding and resolution of the problem and causes. This hierarchy includes screening diagnostic and predictive techniques, which are described.

d. Reservoirs as detention basins. Reservoir mixing is a continual process where low inflows of poor quality are stored and mixed with higher inflows of better quality. Generally, this is accomplished in large reservoirs where annual or even multiple-year flows are retained, but the concept extends to small reservoirs in which weekly or even daily quality changes occur due to variability of loading associated with the inflow.

(1) The use of a reservoir as a mixing devise should be considered whenever the inorganic water quality is unacceptable during some periods but where the average quality falls within the acceptance level. Lake Texoma on the Red River is an example of a reservoir which modifies the quality pattern. Although monthly inflow quality has equalled 1,950 mg/l chloride concentration, the outflow has not exceeded 520 mg/l.

(2) Many materials which enter a reservoir are removed by settling. This applies not only to incoming settleable solids, but also to colloidal and dissolved materials which become of settleable size by chemical precipitation or by synthesis into biological organisms. Reservoirs are often used to prevent such settleable material from entering navigable rivers where settleable materials would interfere with desired uses. However,

reservoirs that receive substantial sediment will have a short useful life. Planning should include evaluation of the ultimate fate and possible replacement of such reservoirs. Reservoir sedimentation is covered in EM 1110-2-4000.

e. Reservoirs as stratified systems. Reservoirs become stratified if density variations caused by temperature or dissolved solids are sufficiently pronounced to prevent complete mixing. This stratification may be helpful or harmful depending on the outlet works, inflow water quality, and the operating procedure of the reservoir.

(1) Temperature stratification can be beneficial for cold-water fisheries if the water which enters the reservoir during the cooler months can also be stored and released during the warmer months. The cooler water released during the warm months can also be valuable as a cooling water source, can provide for higher oxygen transfer (re-aeration) or slower organic waste oxidation (deoxygenation), and can make the water more aesthetically acceptable for water supply and recreational purposes.

(2) Dissolved oxygen stratification usually occurs in density stratified lakes, particularly during the warmer months. The phenomenon occurs because oxygen which has been introduced into the epilimnion by surface re-aeration does not transfer through the metalimnion into the hypolimnion at a rate high enough to satisfy the oxygen demand by dissolved and suspended materials and by the benthal organisms. Thus, the cool bottom waters which are sometimes desirable may be undesirable from a dissolved oxygen standpoint unless energy dissipation structures are constructed to transfer substantial oxygen into the reservoir pool or the reservoir discharge. Mechanical reservoir mixing to equalize temperature and transfer oxygen to lower reservoir levels is one possible tool for managing reservoir water quality.

f. Reservoirs as flow management devices. Reservoirs may improve water quality by merely permitting the management of flow. This management may include maintenance of minimum flows, blending selective releases from one or more reservoirs to maintain a given stream quality, and the exclusion of a flow from a system by diversion.

(1) Minimum flow is often maintained in a stream for navigation, recreation, fish and wildlife, and water rights purposes. Such flows may also aid in maintaining acceptable water quality.

(2) There is general agreement that water may be stored and selectively released to help reduce natural water quality problems where source control is not possible, and

also that water should not be stored and released solely to improve water quality where similar improvement may be achieved by treatment at the source. The use of a water resource to dilute treatable waste materials is regarded as the misuse of a valuable resource in most cases.

(3) Selective release of water from one or more reservoirs may help improve quality at one or more downstream locations. Such releases may be one of the governing factors in establishing reservoir management rules. The water to be released may either be good quality water that

will improve the river quality or poor quality water that is to be discharged when it will do a minimum of harm (e.g., during high flow). The water quality version of the HEC-5 reservoir simulation program (HEC-5Q) is designed to perform quality analysis based on a reservoir simulation for quantity demands and subsequently determine additional releases to meet quality objectives (HEC 1986). Section III, Chapter 4 of EM 1110-2-1201 describes various predictive techniques, including numerical and physical models.

Chapter 12
Conservation Storage Yield

12-1. Introduction

a. Purpose. There are three purposes of this chapter: (1) to provide a descriptive summary of the technical procedures used in the hydrologic studies to analyze reservoirs for conservation purposes; (2) to furnish background information concerning the data requirements, advantages, and limitations of the various procedures; and (3) to establish guidelines which will be helpful in selecting a procedure, conducting the studies, and evaluating results.

b. Procedures. The procedures presented are generally used to determine the relationship between reservoir storage capacity and reservoir yield (supply) for a single reservoir. The procedures may be used to determine storage requirements for water supply, water quality control, hydroelectric power, navigation, irrigation, and other conservation purposes. Although the discussions are limited to single reservoir analysis, many of the principles are generally applicable to multi-reservoir systems. Chapter 4, "Reservoir Systems," presents concepts regarding the analysis of a multi-reservoir system.

12-2. Problem Description

a. Determining storage yield relationships. The determination of storage-yield relationships for a reservoir project is one of the basic hydrologic analyses for reservoirs. The basic objective can be to determine the reservoir yield given a storage allocation, or find the storage required to obtain a desired yield. The determination follows a traditional engineering approach:

(1) Determine the study objectives. This includes the project purposes, operation goals, and the evaluation criteria.

(2) Determine the physical and hydrologic constraints for the site. This includes the reservoir storage and outflow capability, as well as the downstream channel system.

(3) Compile the basic data. The basic data include demands, flow, and losses. Also, the appropriate time interval for analysis, which depends on the data and its variations in time.

(4) Select the appropriate method, one that meets the study objectives and provides reliable information to evaluate results based on accepted criteria.

(5) Perform the analysis, evaluate the results, and present the information.

b. Evaluating hydrologic aspects of planning. Many of the methods described in other chapters of this manual are necessary to develop and provide data to evaluate the hydrologic aspects of reservoir planning, design, and operation. In many cases, the methods required to provide data for a reservoir analysis are more complex than the method for the reservoir study itself. However, because the usefulness and validity of the reservoir analysis are directly dependent upon the accuracy and soundness of basic data, complex methods can often be justified to develop the data.

c. General information. This chapter contains information on types of procedures, considerations of time interval, storage allocation, project purposes, several types of studies, and a summary of methods to analyze the results of reservoir studies. The methods can be characterized as simplified, including sequential and nonsequential analysis, and detailed sequential analysis. Emphasis is given to the sequential routing because:

(1) It is adaptable to study single or multiple reservoir systems.

(2) With an appropriate time interval, the variations in supply and demand can be directly analyzed.

(3) It gives results that are easily understood and explained by engineers familiar with basic hydrologic principles.

(4) The accuracy and results of the study can be closely controlled by the engineers performing and supervising the studies.

(5) It can be used with sparse basic data for preliminary analyses, as well as with detailed data and analyses.

12-3. Study Objectives

a. Establish and consider objectives. Before any meaningful storage-yield analysis can be made, it is necessary to establish and consider the objectives of the hydrologic study. The objectives could range from a preliminary study to a detailed analysis for coordinating reservoir operation for several purposes. The objectives, together with the available data, will control the degree of accuracy required for the study.

b. Determination of storage required. Basically, there are two ways of viewing the storage-yield relationship. The most common viewpoint involves the determination of the storage required at a given site to supply a given yield. This type of problem is usually encountered in the planning and early design phases of a water resources development study.

c. Determination of yield. The second viewpoint requires the determination of yield from a given amount of storage. This often occurs in the final design phases or in re-evaluation of an existing project for a more comprehensive analysis. Because a higher degree of accuracy is desirable in such studies, detailed sequential routings are usually used.

d. Other objectives. Other objectives of a storage-yield analysis include the following: determination of complementary or competitive aspects of multiple project development, determination of complementary or competitive aspects of multiple purpose development in a single project, and analysis of alternative operation rules for a project or group of projects. Each objective and the basis for evaluation dictates implicitly the method which should be used in the analysis.

12-4. Types of Procedures

a. Selecting. The procedures used to determine the storage-yield relationship for a potential dam site may be divided into either simplified or detailed sequential analysis. The selection of the appropriate technical procedure may be governed by the availability of data, study objectives, or budgetary considerations. In general, the simplified techniques are only satisfactory when the study objectives are limited to preliminary or screening studies. Detailed methods are usually required when the study objectives advance to the feasibility and design phases.

b. Simulation and mathematical programming analyses. The detailed sequential methods may be further subdivided into simulation analyses and mathematical programming analyses. In simulation analysis, the physical system is simulated by performing a sequential reservoir routing with specified demands and supply. In this type of study, attempts are made to accurately reproduce the temporal and spatial variation in streamflow and reservoir storage in a reservoir-river system. This is accomplished by accounting for as many significant accretions and depletions as possible. In mathematical programming analysis, the objective is to develop a mathematical model which can be used to analyze the physical system without necessarily reproducing detailed factors. The model usually provides a simulation that will provide a maximum (or minimum) value of the objective function, subject to system constraints.

c. Simplified method. A simplified method can be used if demands for water are relatively simple (constant) or if approximate results are sufficient, as in the case of many preliminary studies. However, it should be emphasized that the objective of the simplified methods is to obtain a good estimate of the results which could be achieved by detailed sequential analysis. Simplified methods consist generally of mass curve and depth duration analyses, which are discussed later.

d. Computer models. Computer models have changed the role of the simplified methods because of the relatively low cost of a detailed sequential routing. Computer programs like HEC-5 *Simulation of Flood Control and Conservation Systems* (HEC 1982c) provide efficient models of reservoirs, based on the level of data availability. For preliminary studies, minimum reservoir and demand information are sufficient. The critical and more complex problem is the development of a consistent flow sequence, which is required by all methods of analysis. Simplified methods still have a role in screening studies or as tools to obtain good estimates of input data for the sequential routings.

e. Detailed sequential routing. In the past, detailed sequential routings have been used almost exclusively for the development of operating plans for existing reservoirs and reservoir systems. However, the advent of the comprehensive basin planning concept, the growing demand for more efficient utilization of water resources, and the increasing competition for water among various project purposes indicate a need for detailed sequential routings in planning studies. Also, these complex system studies provide an opportunity to use optimization to suggest system operations or allocations (ETL 1110-2-336).

f. Mathematical programming. Mathematical programming (optimization) has generally been applied in water management studies of existing systems. The questions addressed usually deal with obtaining maximum gain from available resources, e.g., energy production from a hydropower system. Recent Corps applications include the review of operation plans for reservoir systems, e.g., Columbia and Missouri River Systems (HEC 1991d 1991f, and 1992a). These studies utilized the HEC Prescriptive Reservoir Model HEC-PRM (HEC 1991a).

g. Further information. Wurbs (1991) provides a review of modeling and analysis approaches including simulation models, yield analysis, stochastic streamflow models, impacts of basinwide management on yield, and optimization techniques.

12-5. Factors Affecting Selection

a. Examining objectives and data availability. Before initiating a storage-yield study, the study objectives and data availability should be examined in order to ascertain: (1) the method best suited for the study requirements; (2) the degree of accuracy required to produce results consistent with the study objectives; and (3) the basic data required to obtain the desired accuracy using the selected method. In preliminary studies, limitations in time and scope might dictate the data and method to be used and the accuracy. More detailed analyses are needed when a higher degree of accuracy is desired. A technical study work plan is very useful in organizing study objectives, inventory of available data, and the selection of general procedures.

b. Availability of data. The availability of basic physical and hydrologic data will quite frequently be a controlling factor in determining which of the several technical methods can be used. Obviously, the detailed methods require more data which may not be available. However, detailed simulation can be performed with limited system data if the historic flow data are available. The simplified methods require less data, but the reliability of the results decreases rapidly as the length of hydrologic record decreases. Therefore, it is often desirable to simulate additional hydrologic data for use with simplified methods. Hydrologic data and data simulation are discussed in Chapter 5.

c. Significant aspects. The study level and available data are not the only deciding factors. The study methods must capture the significant aspects of the prototype system. If simplified methods do not utilize data which have major influences upon the results, it would be necessary to utilize a more detailed method which accounts for variations in these data. For example, the National Hydropower Study (USACE 1979) used flow-duration techniques for most reservoirs, but used sequential routing for reservoirs with significant storage.

12-6. Time Interval

a. Selection. The selection of an appropriate time interval depends primarily on the type of analysis and the significance of the data variation over time. Time intervals of one month are usually adequate for nonsequential and preliminary sequential analyses. For more detailed studies, shorter routing intervals will ordinarily be required. Average daily flow data are increasingly used because they are readily available, and computer speed is sufficient to process the data in a reasonable time. Only in exceptional cases will routing intervals of less than one day be required

for conservation studies. Considerable work is involved with shorter intervals, and the effects of time translation, which are usually ignored in conservation routing studies, become important with shorter intervals. Shorter intervals are necessary and should be used during flood periods or during periods when daily power fluctuations occur.

b. Using short time intervals. Ordinarily, when using short time intervals of one day or less, it is necessary to obtain adequate definition of the conditions under study. Periods selected for analysis should exhibit critical combinations of hydrologic conditions and demand characteristics. For example, analysis of hourly power generation at a hydroelectric plant under peaking conditions might be studied for a one-month period where extremely low flows could be assumed to coincide with extremely high power demands. As a rule, studies involving short time intervals are supplementary to one or more studies of longer periods using longer time intervals. The results of the long period study are often used to establish initial conditions such as initial reservoir storage for the selected periods of short-interval analysis.

c. Selecting a routing interval. In sequential conservation routing studies, the selection of a routing interval is dependent upon four major factors: (1) the demand schedule that will be utilized in determining the yield; (2) the accuracy required by the study objectives; (3) the data available for use in the study; and (4) the phase relationship between periods of high and low demands and high and low flow. If the water demand schedule is relatively uniform, it is ordinarily possible to estimate the amount of storage required for a specified yield by graphical analysis using the Rippl diagram or the nonsequential analysis discussed later herein. Demand schedules which show marked seasonal variations usually preclude the use of graphical techniques alone in determining storage requirements. This is especially true when the demand is a function which cannot be described in terms of a specific amount of water, as in the cases of hydroelectric power and water quality. In order to obtain accurate estimates of storage requirements when the demand schedule is variable, it is necessary to make sequential routing studies with routing intervals short enough to delineate important variations in the demand schedule. Simplified techniques may be utilized in obtaining a first estimate of storage requirements for the detailed sequential routing.

d. More accurate results. As a general rule, shorter routing intervals will provide more accurate results. This is due to many factors, such as better definition of relationships between inflow and releases, and better estimates of average reservoir levels for evaporation and power-head calculations. Average flow for longer routing intervals tends to reduce the characteristic variations of streamflows, thus producing a "dampened" storage requirement. This will tend to overestimate yield, or underestimate required storage for reservoirs with small storage. Therefore, the volume of conservation storage, compared to the average flow in a time period, is an additional consideration in selecting the time interval. For example, if a monthly interval is used and there is no sufficient conservation storage to control the variation of flow within the month, the use of average monthly will conceal that fact.

e. Effects on storage requirements. When fluctuations in streamflows or demands have a significant effect on storage requirements, computations should be refined for critical portions of the studies, or shorter routing intervals should be used. However, the routing interval should not be shorter than the shortest period for which flow and demand data are available. Attempts to "manufacture" flow or demand data are usually time consuming and may create errors or give a false impression of accuracy unless reliable information is available for subdivision of basic data.

f. Nonsequential methods. The selection of the flow interval for analysis by nonsequential methods is usually not as critical as for a sequential analysis. Because the nonsequential analysis is restricted to uniform demands, it does not produce results as accurate as those obtained by sequential methods. Therefore, there is very little gain in accuracy with short intervals. Flow intervals of one month are usually suitable for nonsequential methods.

12-7. Physical Constraints

Physical constraints which should be considered in storage-yield studies include conservation storage available, minimum pool, outlet capacities, and channel capacities. The addition of hydropower as a purpose will require the inclusion of constraints to power generation, e.g., maximum and minimum head, penstock capacity, and power capacity. If flood control is to be included as a project purpose, the maximum conservation storage feasible at a given site will be affected by the flood-control analysis.

12-8. Priorities

In order to determine optimum yield in a multiple-purpose project, some type of priority system for the various purposes must be established. This is necessary when the competitive aspects of water use require a firm basis for an operating decision. Safety of downstream inhabitants and cities are of utmost importance, which makes flood reduction the highest priority in a multiple-purpose project during actual operation. During periods of flood operations, conservation requirements might be reduced in order to provide the best flood operation. Although this chapter is not concerned directly with flood-control operation or criteria, it is necessary to integrate flood-control constraints with the conservation study to ensure that operating conditions and reservoir levels for conservation purposes do not interfere with flood-control operation. Priorities among the various conservation purposes vary with locale, water rights, and with the need for various types of water utilization. In multipurpose projects, every effort should be made to develop operation criteria which maximize the complementary uses for the various conservation purposes.

12-9. Storage Limitations

One of the reasons for making sequential conservation routing studies is to determine the effect of storage limitations on yield rates. Simplified yield methods cannot account for operational restrictions imposed by storage limitations in a multiple-purpose project. As shown in Figure 2-2, three primary storage zones, any or all of which may exist in a given reservoir project, may generally be described as follows:

- Exclusive capacity, generally for flood control, in the uppermost storage space in the reservoir.

- Multiple-purpose capacity, typically conservation storage, immediately below the flood control storage.

- Inactive capacity, or dead storage, the lowest storage space in the reservoir.

An additional space, called surcharge, exists between the top of the flood-control space and the top of the dam. Surcharge storage is required to pass flood waters over the spillway. The boundaries between the storage zones and operational boundaries within the zones may be fixed throughout the year, or they may vary from season to season as shown on Figure 12-1. The varying boundaries usually offer a more flexible operational plan which may result in higher yields for all purposes, although an additional element of chance is often introduced when the boundaries are allowed to vary. The purpose of detailed sequential routing studies is to produce an operating scheme and boundary arrangement which minimizes the chance of failure to satisfy any project purpose while optimizing the yield for each purpose. The three storage zones and the effect of varying their boundaries are discussed in the following paragraphs.

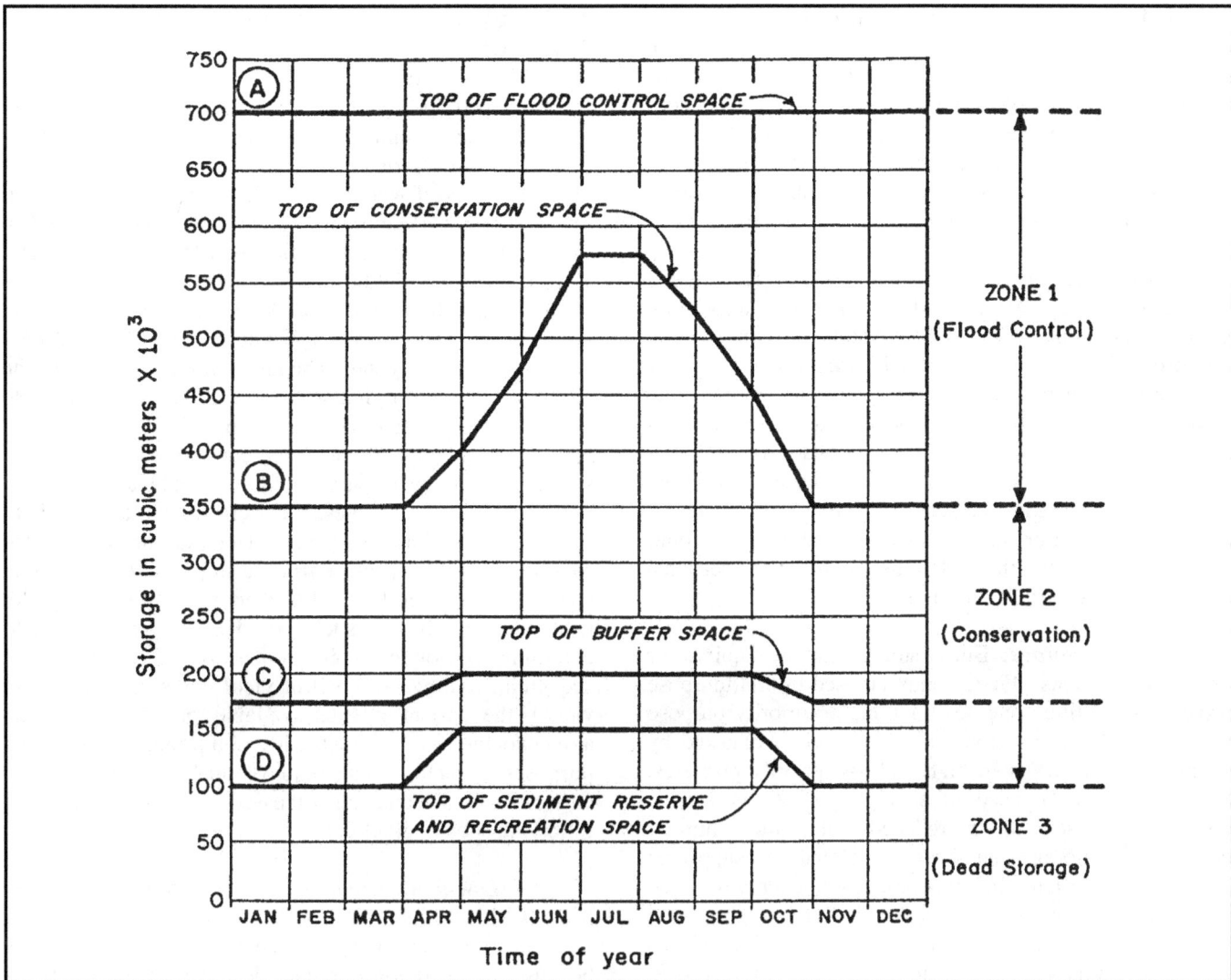

Figure 12-1. Seasonally varying storage boundaries

a. Flood-control storage. The inclusion of flood-control storage in a multiple-purpose project may adversely affect conservation purposes in two ways. First, storage space which might otherwise be utilized for conservation purposes is reserved for flood-control usage. Second, flood-control operations may conflict with conservation goals, with a resultant reduction or loss of conservation benefits. However, detailed planning and analysis of criteria for flood-control and conservation operations can minimize such adverse effects. Even without dedicated flood storage, conservation projects must be able to perform during flood events.

(1) Where competition between flood-control and conservation requirements exists, but does not coincide in time, the use of a seasonally varying boundary between flood-control storage and conservation storage may be used. The general procedure is to hold the top of the conservation pool at a low level when conservation demands are not critical in order to reserve more storage space for flood-control regulation. Then, as the likelihood of flood occurrence decreases, the top of the conservation pool is raised to increase the storage available for conservation purposes. Water management criteria are then tested by detailed sequential routing for the period of recorded streamflow. Several alternative patterns and magnitudes of seasonal variations should be evaluated to determine the response of the storage-yield relationship and the flood-control efficiency to the seasonal variation of the boundary. A properly designed seasonally-varying storage boundary should not reduce the effectiveness of flood-control storage to increase the conservation yield.

(2) Flood-control operation is generally simplified in conservation studies because the routing interval for such studies is frequently too long to adequately define the flood-control operation. Nevertheless, flood-control constraints should be observed insofar as possible. For example, channel capacities below the reservoir are considered for release purposes, and storage above the top of flood-control pool is not utilized.

b. Conservation storage. The conservation storage may be used to regulate minor floods in some multipurpose projects, as well as to supply water for conservation purposes. In addition to seasonal variations in its upper boundary between flood control and conservation, the lower conservation storage boundary may also vary seasonally. If several conservation purposes of different priorities exist, there may be need for a buffer zone in the conservation storage. The seasonal variation in the boundary between conservation storage and buffer zone would be determined by the relationship between seasonal demands for the various purposes.

c. Buffer storage. Buffer storage may be required for one or two reasons. First, it may be used in multipurpose projects to continue releases for a higher priority purpose when normal conservation storage has been exhausted by supplying water for both high and low priority purposes. Second, it may be used in a single-purpose project to continue releases at a reduced rate after normal conservation storage has been exhausted by supplying water at a higher rate. In either case, the boundary between the normal conservation storage and buffer storage is used to change the operational criteria. The location of this boundary and its seasonal variation are important factors in detailed sequential routing because of this change in water management criteria. The amount of buffer storage and, consequently, the location of the seasonally varying boundary between the buffer zone and the remainder of the conservation storage is usually determined by successive approximations in sequential routing studies. However, a simplified procedure, which produces a satisfactory estimate in cases without seasonally varying boundaries, is described in Section 12-11.

d. Inactive or dead storage. The inactive storage zone is maintained in the reservoir for several purposes, such as a reserve for sedimentation or for fish and wildlife habitat. As a rule, the reservoir may not be drawn below the top of the inactive storage. Although it may be possible to vary the top of the reserve pool as shown in Figure 12-1, it is seldom practical to do so. This could reflect the desire to maintain a higher recreation pool during the summer.

12-10. Effects of Conservation and Other Purposes

As previously indicated, the seasonal variation of demand schedules may assume an important role in the determination of required yield. The effect of seasonal variation is most pronounced when the varying demand is large with respect to other demands, as is often the case when hydroelectric power or irrigation is a large demand item. The quantity of yield from a specified storage may be overestimated by as much as 30 percent when a uniform yield rate is used in lieu of a known variable yield rate. Also, variable demand schedules often complicate the analysis of reservoir yield to the extent that it is impossible to accurately estimate the maximum yield or the optimum operation by approximate methods. Because detailed sequential routing is particularly adaptable to the use of variable demand schedules, every effort should be made to incorporate all known demand data into the criteria for routing. Thus, successive trials using detailed sequential analysis must often be used to determine maximum yield. Computer programs such as HEC-5 provide yield determination for reservoirs by performing the successive sequential routing until a firm yield is determined. *Firm yield* is the amount of water available for a specific use on a dependable basis during the life of a project. The project purposes which often require analysis of seasonal variations in demand are discussed in more detail in the following subparagraphs.

a. Low-flow regulation. The operation of a reservoir for low flow regulation at a downstream control point is difficult to evaluate without a detailed sequential routing, because the operation is highly dependent upon the flows which occur between the reservoir and the control point, called intervening local flow. As these flows can vary significantly, a yield based on long period average intervening flows can be subject to considerable error. A detailed sequential routing, in which allowance is made for variations of intervening flows within the routing interval, produces a more reliable estimate of storage requirements for a specified yield and reduces the chance of overestimating a firm yield. Ordinarily, the yield and the corresponding operation of a reservoir for low-flow regulation are determined by detailed sequential routing of the critical period and several other periods of low flow. The entire period of recorded streamflow may not be required, unless summary-type information is needed for functions such as power.

b. Diversion and return flows. The analysis of yield for diversions is complicated by the fact that diversion requirements may vary from year to year as well as from season to season. Furthermore, the diversion requirements may be stated as a function of the natural flow and water rights rather than as a fixed amount. In addition, diversion amounts may often be reduced or eliminated when storage in the reservoir reaches a certain critical low value. When any one of these three items is important to a given reservoir analysis, detailed sequential analysis for the entire period of flow record should be made to determine accurately the yield and the water management criteria. Coordination of the water management criteria for other purposes with the diversion requirements may also be achieved with the detailed sequential analysis results.

c. *Water quality control.* Inclusion of water quality control and management as a project purpose almost always dictates that sequential routing studies be used to evaluate project performance. Practically every variable under consideration in a water quality study will vary seasonally. Following are the variables which must be considered in a water quality study: (1) variation in quality of the inflow, (2) subsequent change in quality of the reservoir waters due to inflow quality and evaporation, (3) variation in quality of natural streamflow entering the stream between the reservoir and the control station, (4) variation in effluent from treatment plants and storm drainage outflow between the reservoir and the control station, and (5) variation in quality requirements at the control station. Accurate evaluation of project performance must consider all of these variations which pertain to water quality control. Additionally, there are several quality parameters which may require study, and each parameter introduces additional variations which should be evaluated. For example, if temperature is an important parameter, the level of the reservoir from which water is released should be considered in addition to the above variables. Likewise, if oxygen content is important, the effects of release through power units versus release through conduits must be evaluated.

d. *Hydroelectric power generation.*

(1) If hydroelectric power is included as a project purpose, detailed sequential routings are necessary to develop water management criteria, to coordinate power production with other project purposes, and to determine the project's power potential. As a rule, simplified methods are usable for power projects only for preliminary or screening studies, reservoirs with very little power storage, or when energy is a by-product to other operations. Flow-duration analysis, described in Chapter 11, is typically applied in these situations. Power production is a function of both head and flow, which requires a detailed sequential study when the conservation storage is relatively large and the head can be expected to fluctuate significantly.

(2) Determination of firm power or firm energy is usually based on sequential routings over the critical period. The critical period must consider the combination of power demand and critical hydrologic conditions. Various operational plans are used in an attempt to maximize power output while meeting necessary commitments for other project purposes. When the optimum output is achieved, a water management guide curve can be developed. The curve is based on the power output itself and on the plan of operation followed to obtain the maximum output. Critical period analysis and curve development are described in Section 12-11. Additional sequential routings for the entire period of flow record are then made using the rule curve developed in the critical period studies. These routings are used to coordinate power production with flood-control operation and to determine the average annual potential energy available from the project.

(3) In areas where hydroelectric power is used primarily for peaking purposes, it is important that storage requirements be defined as accurately as possible because the available head during a period of peak demand is required to determine the peaking capability of the project. An error in storage requirements, on the other hand, can adversely affect the head with a resultant loss of peaking capability.

(4) Tailwater elevations are also of considerable importance in power studies because of the effect of head on power output. Several factors which may adversely affect the tailwater elevation at a reservoir are construction of a reregulation reservoir below the project under consideration, high pool elevations at a project immediately downstream from the project under consideration, and backwater effects from another stream if the project is near the confluence of two streams. If any of these conditions exist, the resultant tailwater conditions should be carefully evaluated. For projects in which peaking operation is anticipated, an assumed "block-loading" tailwater should be used to determine reservoir releases for the sequential reservoir routing. The "block-loading" tailwater elevation is defined as the tailwater elevation resulting from sustained generation at or near the plant's rated capacity which represents the condition under which the project is expected to operate most of the time. Although in reality the peaking operation tailwater would fluctuate considerably, the use of block-loading tailwater elevation ensures a conservative estimate of storage requirements and available head.

(5) Reversible pump-turbines have enhanced the feasibility of the pumped-storage type of hydroelectric development. Pumped storage includes reversible pump-turbines in the powerhouse along with conventional

generating units, and an afterbay is constructed below the main dam to retain water for pumping during nongenerating periods. Sequential routing studies are required for analyses of this type because of the need to define storage requirements in the afterbay, pumping requirements and characteristics, and the extent to which plan should be developed. Many of the existing and proposed pumped-storage projects in the United States, however, are single purpose projects which do not have conventional units and often utilize off-channel forebays.

12-11. Simplified Methods

If demands for water are relatively constant or if approximate results are sufficient, as in the case of many preliminary studies, a simplified method can be used to save time and effort. The use of simplified techniques which do not consider sequential variations in streamflow or demand are generally limited to screening studies or developing first estimates of storage or yield. The following procedures will generally produce satisfactory results and continue to have a role in storage-yield determinations.

a. Sequential mass curve. The most commonly used simplified sequential method is the sequential mass curve analysis, sometimes referred to as the Rippl Method. This method produces a graphical estimate of the storage required to produce a given yield, assuming that the seasonal variations in demand are not significant enough to prohibit the use of a uniform draft (demand) rate. The sequential mass curve is constructed by accumulating inflows to a reservoir site throughout the period of record and plotting these accumulated inflows versus the sequential time period as illustrated in Figure 12-2.

(1) The desired yield rate, in this example 38,000 m³/year, is represented by the slope of a straight line. Straight lines are then constructed parallel to the desired yield rate and tangent to the mass curve at each low point (line ABC) and at the preceding high point that gives the highest tangent (line DEF). The vertical distance between these two lines (line BE) represents the storage required to provide the desired yield during the time period between the two tangent points (points D and B). The maximum vertical difference in the period is the required

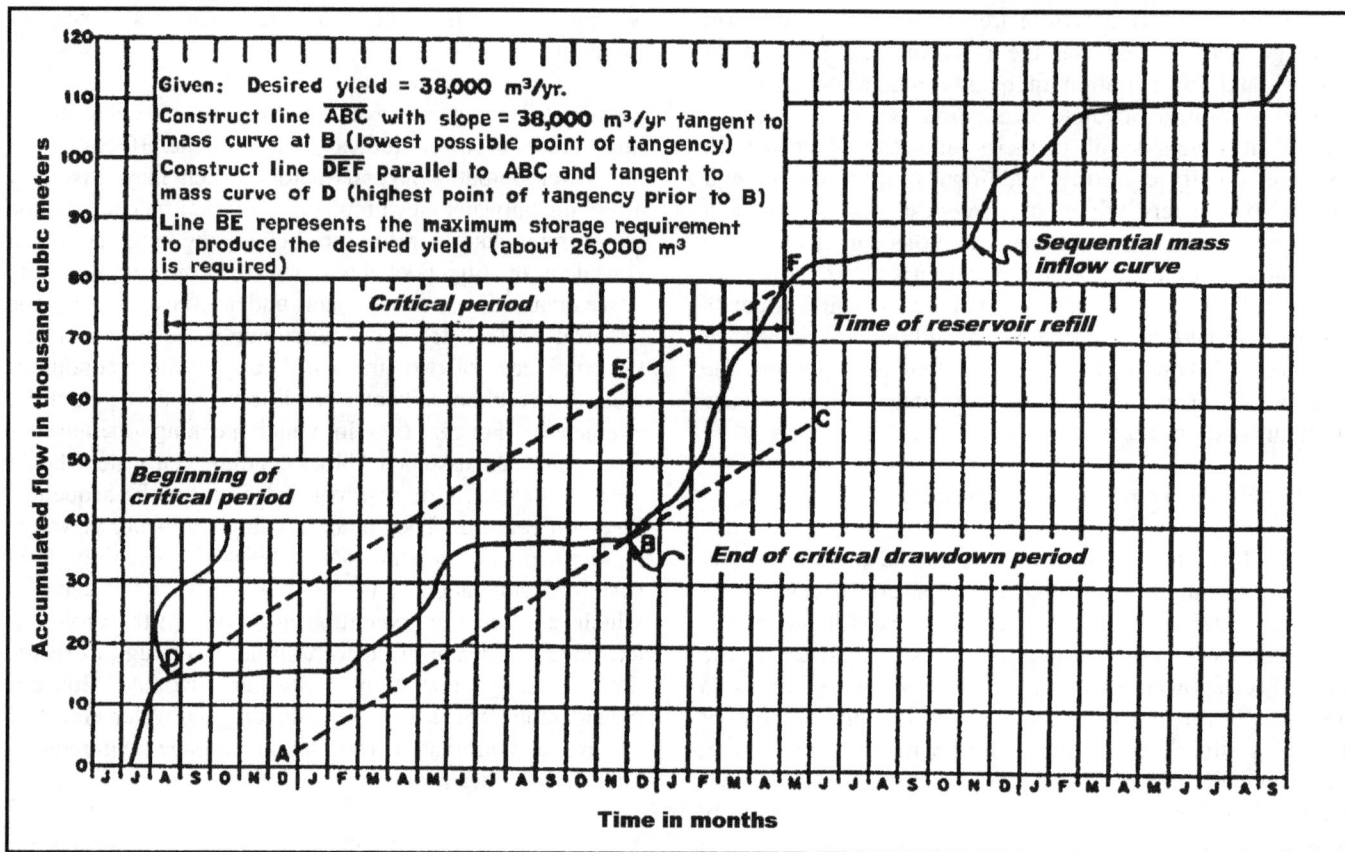

Figure 12-2. Storage determination using a sequential mass curve

storage to meet the desired yield, during the given flow sequence.

(2) The critical period is the duration of time from point **D**, when conservation storage drawdown begins, to point **F**, when the reservoir conservation storage fills. The critical drawdown is from point **D** to point **B**, while during the time from **B** to **F** the reservoir would be refilling.

(3) The sequential mass curve method does not indicate the relative frequency of a shortage. However, by using nonsequential methods, a curve of yield versus shortage frequency can be determined.

b. Nonsequential mass curve. Several nonsequential methods can be used to develop a relation for storage yield versus shortage frequency. The application of this procedure is limited, however, to water supply demands that are uniform in time. These methods involve the development of probability relations for varying durations of streamflow. The historical flows, supplemented by simulated flows where needed, are used to determine frequency tables of independent low-flow events for several durations. A series of low-flow events for a particular duration is selected by computing and arranging in order of magnitude, the independent minimum-flow rates for that duration for the period of record.

(1) After the frequency tables of independent low-flow events are computed for various durations, low-flow frequency curves are obtained by plotting the average flow on log-probability paper. Chapter 4 of EM 1110-2-1415 describes the procedure and presents an example table and frequency plot.

(2) Care must be exercised in the interpretation of the low-flow curves because the abscissa is "nonexceedance frequency per 100-years," or the number of events within 100-years that have a flow equal to or less than the indicated flow. Thus, when low-flow durations in excess of one year are evaluated, the terminology cannot be used interchangeably with probability. For instance, during a 100-year period, the maximum number of independent events of 120 months (10-years) duration is 10. Therefore, the 120-month duration curve cannot cross the value of 10 on the "nonexceedance frequency per 100-year" scale.

(3) Minimum runoff-duration curves for various frequencies, as shown in Figure 12-3 are obtained by plotting points from the low-flow frequency curves on logarithmic paper. The flow rates are converted to volumes (millions of cubic meters in this example). The logarithmic scales simply permit more accurate interpolation between durations represented by the frequency curves.

(4) The nonsequential mass curve (Figure 12-4) is developed by selecting the desired volume-duration curve from Figure 12-3 and plotting this curve on arithmetic grid. The desired yield is then used to determine the storage requirement for the reservoir. The storage requirement is determined by drawing a straight line, with slope equivalent to the required gross yield, and by plotting this line tangent to the mass curve. The absolute value of the negative vertical intercept represents the storage requirement. The application of this procedure is severely limited everywhere in the case of seasonal variations in runoff and yield requirements because the nonsequential mass curve does not reflect the seasonal variation in streamflows, and the tangent line does not reflect seasonal variations in demand. However, the method does provide an estimate of yield reliability.

c. Evaporation losses. Another disadvantage of these simplified types of storage-yield analysis is the inability to evaluate evaporation losses accurately. This may not be critical in humid areas where net evaporation (lake evaporation minus pre-project evapotranspiration) is relatively small, but it can cause large errors in studies for arid regions. Also, these procedures do not permit consideration of seasonal variations in requirements, system nonlinearities, conflicting and complementing service requirements, and several other factors.

12-12. Detailed Sequential Analysis

a. Sequential analysis. Sequential analysis is currently the most accepted method of determining reservoir storage requirements. Many simplified methods have given way to the more detailed computer simulation approaches. In many instances, the computer solution provides more accurate answers at a lower cost than the simple hand solutions.

b. Accounting for reservoir water. Sequential analysis applies the principle of conservation of mass to account for the water in a reservoir. The fundamental relationship used in the routing can be defined by:

$$I - O = \Delta S \qquad (12\text{-}1)$$

where

I = total inflow during the time period, in volume units

O = total outflow during the time period, in volume units

ΔS = change in storage during the time period, in volume units

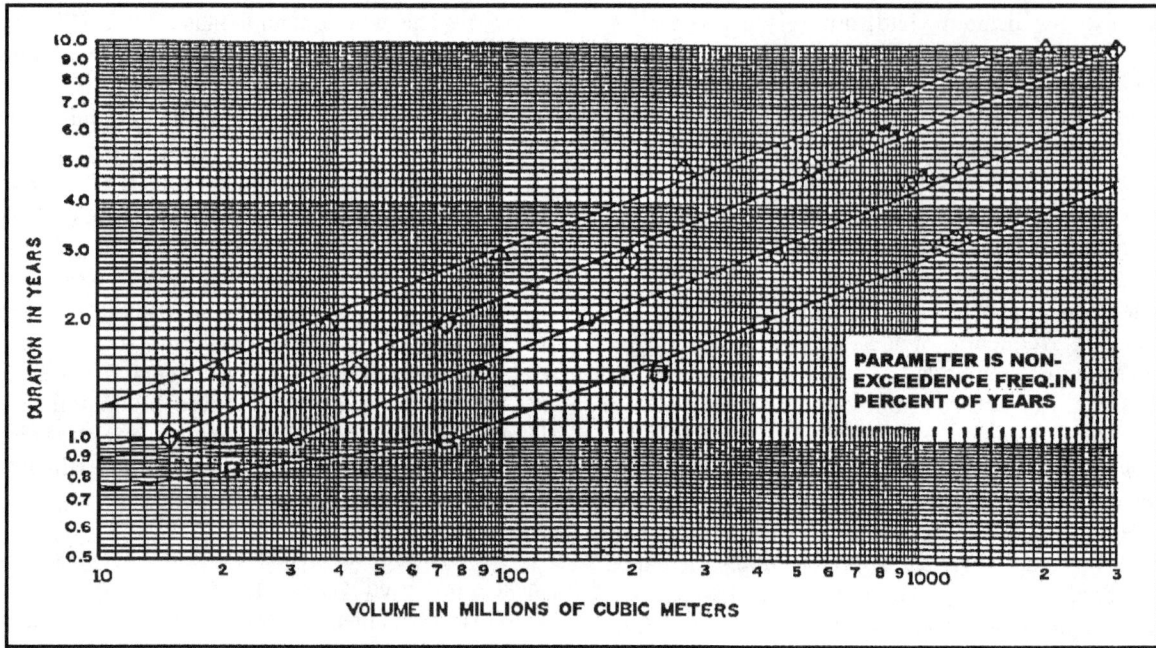

Figure 12-3. Minimum runoff-duration curves

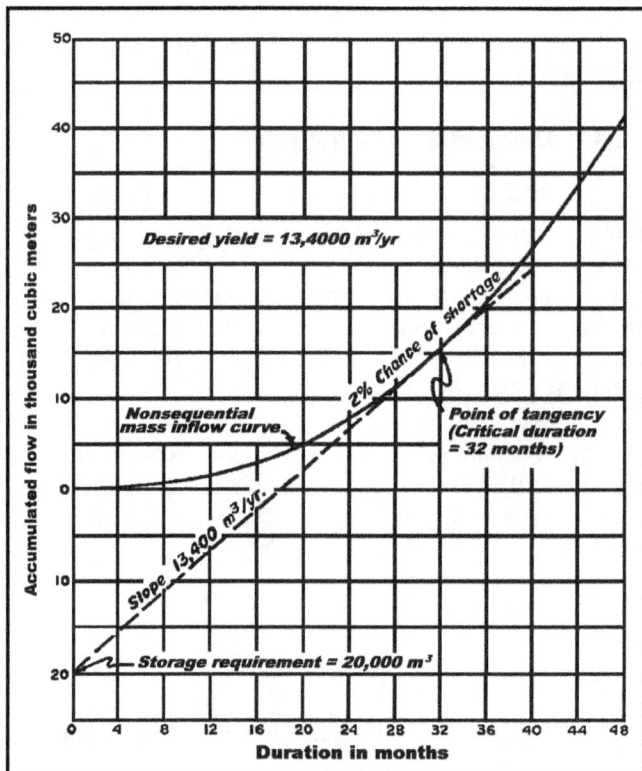

Figure 12-4. Nonsequential mass curve from
Figure 12-3

The inflow and outflow terms include all types of inflow
and outflow. The inflows should include natural streamflow, releases from upstream reservoirs, local inflow to the
reservoir, precipitation falling on the reservoir surface
(sometimes included in computation of net evaporation),
and diversions into the reservoir. Outflows consist of
reservoir releases plus evaporation losses, leakage, and
diversions out of the reservoir. Sequential routing provided the framework for accounting for all water in the
system. The application can be as detailed as required.
The development of the required data constitutes the major
effort in most sequential routing studies.

  c. *Sequential routing.* Sequential routing uses a
repetitive solution of Equation 12-1 in the form of:

$$S_t = S_{t-1} + I_t - O_t \qquad (12\text{-}2)$$

where

  S_t = storage at the end of time t, volume units

  S_{t-1} = storage at the end of time t-1, volume units

I_t = average inflow during time step Δt, converted to volume units

O_t = average outflow during time step Δt, converted to volume units

The primary input includes reservoir storage capacity and allocation, requirements, losses, flow at all model locations for the simulation period, and system connectivity and constraints. The primary output for the reservoir is the average reservoir release for each time step and the resulting reservoir storage at the end of each time step. The releases are made to meet specified requirements, subject to all specified constraints such as storage allocation and maximum release capability. Downstream accounting of flows adds reservoir releases and subtracts diversions and losses to the local downstream flows to compute regulated flow at desired locations. If a short time interval is used, the flow travel time must be considered. Channel routing is usually done with hydrologic routing methods.

d. Multipurpose reservoir routings. The HEC-5 *Simulation of Flood Control and Conservation Systems* (HEC 1982c) computer program performs multipurpose reservoir routings for reservoir systems providing for services at the reservoirs and downstream control points. Releases from a reservoir are determined by the specified requirements for project purposes. Reservoir releases may be controlled at the dam site by hydroelectric power requirements, downstream control for flow, diversion, water rights, or quality. Additional diversions may be made directly from the reservoir.

(1) The program operates to meet the downstream flow requirement, considering available supplies and supplement flow from the intervening area. The storage allocation and most demands can be defined as constant, monthly varying, or seasonally varying. Historic simulations can be performed with period-by-period demands for low flow and hydropower.

(2) Computer program HEC-5 has a firm yield routine called optimization of conservation storage in the program user's manual. The routine can either determine the required storage to meet a specified demand or the maximum reservoir yield that can be obtained from a specified amount of storage. While designed for a single reservoir, it can use up to six reservoirs in a single run, provided the reservoirs operate independently. Optimization can be accomplished on monthly firm energy requirements, minimum monthly flow, monthly diversions, or all of the conservation purposes. The routine can estimate the critical period and make a firm yield estimate based on that

period. After the firm yield is estimated, the program will perform a period of record simulation to ensure that the firm yield can be met. Several cycles of critical period and period of record simulations can be performed in one computer run, based on user input specifications.

12-13. Effects of Water Deficiencies

a. Water storages. Absolute guarantees of water yield are usually not practical, and the designer should therefore provide estimates of shortages that could reasonably develop in supplying the demands with available storage. If nonsequential procedures have been used, information on future shortages is limited to the probability or frequency of occurrence, and the duration or severity of shortages will not be known. In using the Rippl Method, the computations are based on just meeting the demand; therefore, no shortages are allowed during the period of analysis. The result gives no information on the shortages that might be expected in the future. Only in the detailed sequential analysis procedure is adequate information on expected future shortages obtainable.

b. Amount and duration of water shortage. The amount and duration of shortage that can be tolerated in serving various project purposes can greatly influence the amount of storage required to produce a firm yield. These tolerances vary a great deal for different project purposes and should be analyzed carefully in reservoir design. Also, changes in reservoir operation should be considered to meet needs during drought (HEC 1990a).

c. Intolerable shortages. Shortages are generally considered to be intolerable for purposes such as drinking water. However, some reduction in the quantity of municipal and industrial water required can be tolerated without serious economic effects by reducing some of the less important uses of water such as lawn watering, car washing, etc. Shortages greater than 10 percent may cause serious hardship. Most designs of reservoir storage for municipal and industrial water supply are based on supplying the firm yield during the most critical drought of record. Typically, drought contingency plans are developed to meet essential demands during drought conditions that may be more severe than the historic critical period. ETL 1110-2-335 provides guidance for developing and updating plans.

d. Irrigation shortages. For irrigation water supply, shortages are usually acceptable under some conditions. Often the desired quantity can be reduced considerably during the less critical parts of the growing season without great crop loss. Also, if there is a reliable forecast of a drought, the irrigator may be able to switch to a crop having less water requirements or use groundwater to make up the deficit. Shortages of 10 percent usually have negligible economic effect, whereas shortages as large as 50 percent are usually disastrous.

e. Water supply for navigation and low flow augmentation. In designing a reservoir to supply water for navigation or low-flow augmentation, the amount and duration of shortages are usually much more important than the frequency of the shortages. Small shortages might only require rescheduling of fully-loaded vessels, whereas, large shortages might stop traffic altogether. The same thing is true for such purposes as fish and wildlife where one large shortage during the spawning season, for example, could have serious economic effects for years to come.

f. Effects of shortages. Each project purpose should be analyzed carefully to determine what the effects of shortages will be. In many cases, this will be the criterion that determined the ultimate amount of reservoir storage needed for water supply and low-flow regulation.

12-14. Shortage Index

a. Definition. A general approach to shortage definition is to use a shortage index. The shortage index is equal to the sum of the squares of the annual shortages over a 100-year period when each annual shortage is expressed as a ratio to the annual requirements, as shown below:

$$SI = \frac{100 \sum\limits_{i=1}^{i=N} \left[\dfrac{S_A}{D_A} \right]^2}{N} \qquad (12\text{-}3)$$

where

SI = shortage index

N = number of years in routing study

S_A = annual shortage (annual demand volume minus annual volume supplied)

D_A = annual demand volume

This shortage index reflects the observation that economic and social effects of shortages are roughly proportional to the square of the degree of shortage. For example, a shortage of 40 percent is assumed to be four times as severe as a shortage of 20 percent. Similarly, as illustrated in Table 12-1, shortages of 50 percent during 4 out of

Table 12-1
Illustration of Shortage Index

Shortage Index	No. of Annual Shortages per 100 Years	Annual Shortage (S_A/D_A) In %
1.00	100	10
1.00	25	20
1.00	4	50
0.25	25	10
0.25	1	50

100 years are assumed four times as severe as shortages of 10 percent during 25 out of 100 years.

The shortage index has considerable merit over shortage frequency alone as a measure of severity because shortage frequency considers neither magnitude nor duration. The shortage index can be multiplied by a constant to obtain a rough estimate of associated damages.

b. Additional criteria needed. There is a definite need for additional criteria delineating shortage acceptability for various services under different conditions. These criteria should be based on social and economic costs of shortages in each individual project study, or certain standards could be established for the various services and conditions. Such criteria should account for degree of shortage as well as expected frequency of shortage.

12-15. General Study Procedures

a. Water supply. After alternative plans for one or more water supply reservoirs have been established, the following steps can be followed in performing hydrologic studies required for each plan:

(1) Obtain all available daily and monthly streamflow records that can be used to estimate historical flows at each reservoir and diversion or control point. Compute monthly flows and adjust as necessary for future conditions at each pertinent location. A review of hydrologic data is presented in Chapter 5.

(2) Obtain area-elevation data on each reservoir site to be studied and compute storage capacity curves. Determine maximum practical reservoir stage from physiographic and cultural limitations.

(3) Estimate monthly evapotranspiration losses from each site and monthly lake evaporation that is likely to occur if the reservoir is built.

(4) Determine seasonal patterns of demands and total annual requirements for all project purposes, if applicable,

as a function of future time. Synthesize stochastic variations in demands, if significant.

(5) Establish a tentative plan of operation, considering flood control and reservoir sedimentation as well as conservation requirements, and perform an operation study based on runoff during the critical period of record. The HEC-5 *Simulation of Flood Control and Conservation Systems* computer program can be used for this purpose.

(6) Revise the plan of operation, including sizes of various facilities, as necessary to improve accomplishments and perform a new operation study. Repeat this process until a near-optimum plan of development is obtained.

(7) Depending on the degree of refinement justified in the particular study, test this plan of development using the entire period of estimated historical inflows and as many sequences of synthetic streamflows and demands as might be appropriate. Methods for developing synthetic flow sequences are presented in Chapter 12 of the Hydrologic Frequency manual (EM 1110-2-1415).

(8) Modify the plan of development to balance yields and shortages for the maximum overall accomplishment of all project objectives.

b. Hydroelectric power. The study procedure for planning, designing, and operating hydroelectric developments can be summarized as follows:

(1) From an assessment of the need for power generation facilities, obtain information concerning the feasibility and utility of various types of hydroelectric projects. This assessment could be made as part of the overall study for a given project or system, or it could be available from a national, regional, or local power authority.

(2) From a review of the physical characteristics of a proposed site and a review of other project purposes, if any, develop an estimate of the approximate amount of space that will be available for either sole- or joint-use power storage. This determination and the needs developed under step (1) will determine whether the project will be a storage, run-of-river, or pumped-storage power project and whether it will be operated to supply demands for peaking or for baseload generation.

(3) Using information concerning seasonal variation in power demands obtained from the assessment of needs, and knowing the type of project and the approximate storage usable for power production, determine the historical critical hydro-period by review of the historical hydrologic data.

(4) An estimate of potential hydroelectric energy for the assumed critical hydro-period is made using Equation 11-2. If the energy calculated from this equation is for a period other than the basic marketing contract period (usually a calendar year), the potential energy during the critical hydro-period should be converted to a firm or minimum quantity for the contract period (minimum annual or annual firm in the case of a calendar year).

(5) Because the ability of a project to produce hydroelectric energy and peaking capacity is a complex function of the head, the streamflow, the storage, and operation for all other purposes, the energy estimate obtained in step (4) is only an approximation. Although this approximation is useful for planning purposes it should be verified by simulating the operation of the project for all authorized purposes by means of a sequential routing study. Chapter 11 provides methods for performing and analyzing sequential routing studies.

(6) From the results of detailed sequential routing studies, the data necessary for designing power-generating units and power-related facilities of the project should be developed. The design head and design output of the generating units, approximate powerhouse dimensions, approximate sizes of water passages, and other physical dimensions of the project depend on the power installation.

(7) Operation rules for the project must be developed before construction is completed. These rules are developed and verified through sequential routing studies that incorporate all of the factors known to affect the project's operation. For many multipurpose projects, these operation rules are relatively complex and require the use of computerized simulation models to facilitate the computations involved in the sequential routing studies.

(8) If the project is to be incorporated into an existing system or if the project is part of a planned system, system operation rules must be developed to define the role of the project in supplying energy and water to satisfy the system demands. These rules are also developed and tested using sequential routing studies. Sequential routing studies for planning or operating hydroelectric power systems are best accomplished using a computer program such as HEC-5.

Chapter 13
Reservoir Sedimentation

13-1. Introduction

"The ultimate destiny of all reservoirs is to be filled with sediment," (Linsley et al. 1992). The question is how long will it take? Also, as the sediment accumulates with time, will it adversely affect water control goals?

a. Transport capacity. A reservoir changes the hydraulics of flow by forcing the energy gradient to approach zero. This results in a loss of transport capacity with the resulting deposition. The smaller the particles, the farther they will move into the reservoir before depositing. Some may even pass the dam. Deep reservoirs are not fully mixed and are conducive to the formation of density currents.

b. Sediment deposits. The obvious consequence of sediment deposits is a depletion in reservoir storage capacity. Figure 13-1 illustrates components of sediment deposition in a deep reservoir. The volume of sediment material in the delta and the main reservoir depends on the inflowing water and sediment, reservoir geometry, project operation and life among other things. The delta will continue to develop, with time, and the reservoir will eventually fill with sediment.

13-2. Reservoir Deposition

a. Total available sediment. The first step is to estimate the total sediment that will be available for deposition during the design life of the project. Required data include design life of the reservoir, reservoir capacity, water and sediment yield from the watershed, the composition of the sediment material, and the unit weight of sediment deposits. With this information, the trap efficiency can be determined.

b. Trap efficiency. Trap efficiency is the percent of inflowing sediment that remains in the reservoir. Some proportion of the inflowing sediment leaves the reservoir through the outlet works. The proportion remaining in the reservoir is typically estimated based on the trap efficiency. Trap efficiency is described in Section 3-7(a) of EM 1110-2-4000, and the calculations are described in an appendix therein. The efficiency is primarily dependent on the detention time, with the deposition increasing as the time in storage increases.

c. Existing reservoirs. Existing reservoirs are routinely surveyed to determine sediment deposition, and resulting loss of storage. Section 5-30 and Appendix K of EM 1110-2-4000 describe the Corps program. This historic deposition data can be useful for checking computed estimates. "Sediment Deposition in U.S. Reservoirs (Summary of Data Reported 1981-85)" provides

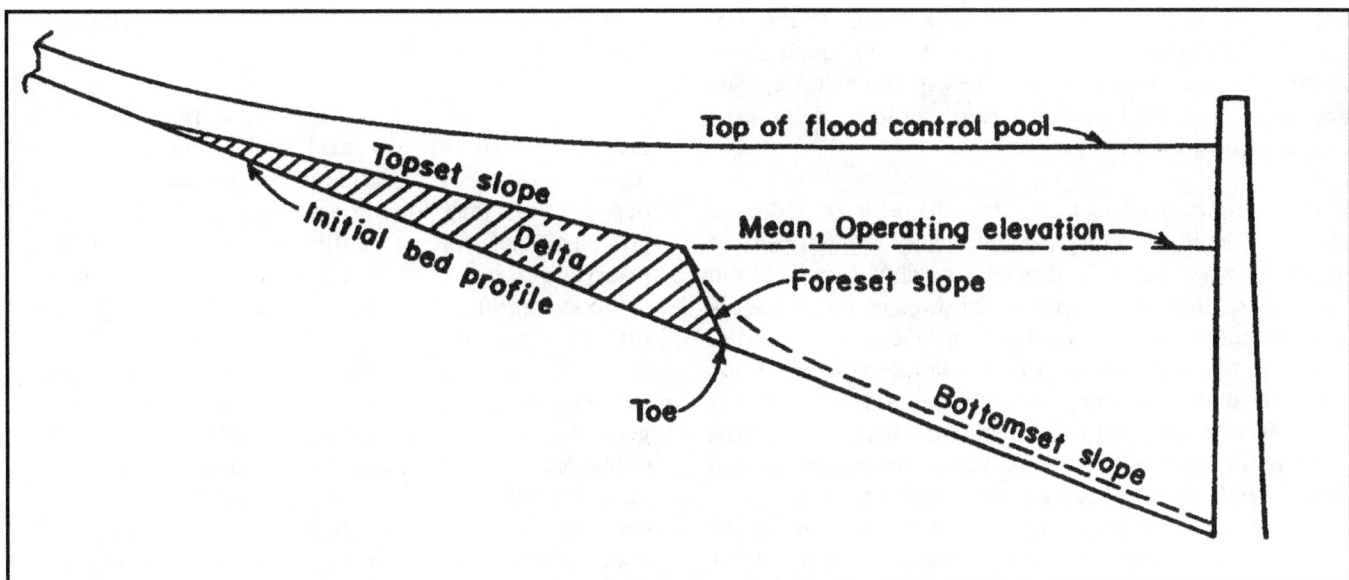

Figure 13-1. Conceptual deposition in deep reservoirs

data on reservoir locations, drainage areas, survey dates, reservoir storage capacities, ratios of reservoir capacities to average annual inflows, specific weights (dry) of sediment deposits, and average annual sediment-accumulation rates (U.S. Geological Survey 1992). Reservoirs are grouped by drainage basins.

13-3. Distribution of Sediment Deposits in the Reservoir

The planning or design of a reservoir requires an analysis to determine how sediment deposits will be distributed in the reservoir. This is a difficult aspect of reservoir sedimentation because of the complex interaction between hydraulics of flow, reservoir operating policy, inflowing sediment load, and changes in the reservoir bed elevation. The traditional approach to analyzing the distribution of deposits has relied on empirical methods, all of which require a great deal of simplification from the actual physical problem.

a. Main channel deposition. Conceptually, deposition starts in the main channel. As flow enters a reservoir, the main channel fills at the upstream end until the elevation is at or above the former overbank elevations on either side. Flow then shifts laterally to one side or the other, but present theory does not predict the exact location. During periods of high water elevation, deposition will move upstream. As the reservoir is drawn down, a channel is cut into the delta deposits and subsequent deposition moves material farther into the reservoir. The lateral location of the channel may shift from year to year, but the hydraulic characteristics will be similar to those of the natural channel existing prior to impounding the reservoir. Vegetation will cover the exposed delta deposits and thus attract additional deposition until the delta takes on characteristics of a floodplain.

b. Sediment diameters. The diameter of sediment particles commonly transported by streams ranges over five log cycles. Generally, the coarse material will settle first in the outer reaches of the reservoir followed by progressively finer fractions farther down toward the reservoir dam. Based on this depositional pattern, the reservoir is divided into three distinct regions: top-set, fore-set, and bottom-set beds. The top-set bed is located in the upper part of the reservoir and is largely composed of coarse material or bed load. While it may have a small effect on the reservoir storage capacity, it could increase upstream stages. The fore-set region represents the live storage capacity of the reservoir and comprises the wash load. The bottom-set region is located immediately upstream of the dam and is

primarily composed of suspended sediments brought from upstream by density currents. The region is called the reservoir dead storage and generally does not affect the storage capacity. Some of the finest material may not settle out and will pass through the dam. In order to calculate the volume of material which will deposit as a function of distance, grain size must be included as well as the magnitude of the water discharge and the operating policy of the reservoir.

c. Reservoir shape. Reservoir shape is an important factor in calculating the deposition profile. For example, flow entering a wide reservoir spreads out, thus reducing transport capacity, but the path of expanding flow does not necessarily follow the reservoir boundaries. It becomes a 2-dimensional problem to calculate the flow distribution across the reservoir in order to approximate transport capacity and, therefore, the resulting deposition pattern. On the other hand, flow entering a narrow reservoir has a more uniform distribution across the section resulting in hydraulic conditions that are better approximated by 1-dimensional hydraulic theory.

d. Flood waves. Flood waves attenuate upon entering a reservoir. Therefore, their sediment transport capacity decreases from two considerations: (1) a decrease in velocity due to the increase in flow area and (2) a decrease in velocity due to a decrease in water discharge resulting from reservoir storage. As reservoir storage is depleted by the sediment deposits in the delta, the impact of attenuation on transport capacity diminishes. The resulting configuration, therefore, is assumed to depend upon the first consideration, whereas, the time for delta development is influenced somewhat by the second consideration.

e. Flood-pool index method. If flood control is a project purpose, the next level of detail in reservoir sedimentation studies is to divide the total volume of predicted deposits into that volume settling into the flood-control pool and that volume settling in the remainder of the reservoir. The flood-pool index method requires the depth of flood-control pool, depth of reservoir, and the percent of time the reservoir water level is at or above the bottom of the flood-control pool. Based on the index, the percent of sediment trapped in the flood-control pool is estimated by a general empirical relationship. Appendix H of EM 1110-2-4000 describes the index method and provides several other methods for estimating the distribution of sediment deposits in reservoirs. Chapter 5, Section IV, EM 1110-2-4000, provides an overview of levels of sedimentation studies and methods of analysis.

PART 4

HYDROLOGIC ENGINEERING STUDIES FOR RESERVOIRS

Chapter 14
Spillways and Outlet Works

14-1. Function of Spillways and Outlet Works

Spillways and outlet works are necessary to provide capability to release an adequate rate of water from the reservoir to satisfy dam safety and water control regulation of the project. Sections 4-2 and 4-3 in EM 1110-2-3600 provide general descriptions of types and operation requirements for spillways and outlet works, respectively.

a. Spillway adequacy. While the outflow capability must be provided throughout the operational range of the reservoir, the focus of hydrologic studies is usually on the high flows and spillway adequacy. Dam failures have been caused by improperly designed spillways or by insufficient spillway capacity. Ample capacity is of great importance for earthfill and rockfill dams, which are likely to be destroyed if overtopped; whereas concrete dams may be able to withstand moderate overtopping.

b. Spillway classification. Spillways are ordinarily classified according to their most prominent feature, either as it pertains to their shape, location, or discharge channel. Spillways are often referred to as controlled or uncontrolled, depending on whether they are gated or ungated. EM 1110-2-1603 describes a variety of spillway types and provides hydraulic principles, design criteria, and results from laboratory and prototype tests.

c. Outlet works. Outlet works serve to regulate or release water impounded by a dam. It may release incoming flows at a reduced rate, as in the case of a detention dam; divert inflows into canals or pipelines, as in the case of a diversion dam; or release stored-water at such rates as may be dictated by downstream needs, evacuation considerations, or a combination of multiple-purpose requirements.

d. Outlet structure classification. Outlet structures can be classified according to their purpose, their physical and structural arrangement, or their hydraulic operation. EM 1110-2-1602 provides information on basic hydraulics, conduits for concrete dams, and conduits for earth dams with emphases on flood-control projects. Appendix IV of EM 1110-2-1602 provides an illustrative example of the computation of a discharge rating for outlet works.

e. Low level outlets. Low level outlets are provided to maintain downstream flows for all levels of the reservoir operational pool. The outlets may also serve to empty the reservoir to permit inspection, to make needed repairs, or to maintain the upstream face of the dam or other structures normally inundated.

f. Outlets as flood-control regulators. Outlet works may act as a flood-control regulator to release waters temporarily stored in flood control storage space or to evacuate storage in anticipation of flood inflows. In this case, the outflow capacity should be able to release channel capacity, or higher. The flood control storage must be evacuated as rapidly as safely possible, in order to maintain flood reduction capability.

14-2. Spillway Design Flood

a. Spillway design flood analyses. Spillway design flood (SDF) analyses are performed to evaluate the adequacy of an existing spillway or to size a spillway. For a major project, the conservative practice in the United States is to base the spillway design flood on the probable maximum precipitation (PMP). The PMP is based on the maximum conceivable combination of unfavorable meteorological events. While a frequency is not normally assigned. a committee of ASCE has suggested that the PMP is perhaps equivalent to a return period of 10,000 years.

b. Probable maximum flood. The PMF inflow hydrograph is developed by centering the PMP over the watershed to produce a maximum flood response. The unit hydrograph approach, described in Chapter 7 of this manual, is usually applied. Section 13-5 of EM 1110-2-1417 contains information on PMP determination and computation of the PMF.

c. Flood hydrographs. The inflow design flood hydrographs are usually for rainfall floods. Normally, such floods will have the highest peak flows but not always the largest volumes. When spillways of small capacities in relation to these inflow design flood peaks are considered, precautions must be taken to ensure that the spillway capacity will be sufficient to evacuate storage so that the dam will not be overtopped by a recurrent storm, and prevent the flood storage from being kept partially full by a prolonged runoff whose peak, although less than the inflow design flood, exceeds the spillway capacity. To meet these requirements, the minimum spillway capacity should be in accord with the following general criteria (Hoffman 1977):

(1) In the case of snow-fed perennial streams, the spillway capacity should never be less than the peak discharge of record that has resulted from snowmelt runoff.

(2) The spillway capacity should provide for the evacuation of sufficient surcharge storage space so that in routing a succeeding flood, the maximum water surface does not exceed that obtained by routing the inflow design flood. In general, the recurrent storm is assumed to begin 4 days after the time of peak outflow obtained in routing the inflow design flood.

(3) In regions having an annual rainfall of 40 in. or more, the time interval to the beginning of the recurrent storm in criterion (2) should be reduced to 2 days.

(4) In regions having an annual rainfall of 20 in. or less, the time interval to the beginning of the recurrent storm in criterion (2) may be increased to 7 days.

14-3. Area and Capacity of the Reservoir

a. Reservoir capacity and operations. Dam designs and reservoir operating criteria are related to the reservoir capacity and anticipated reservoir operations. The reservoir capacity and reservoir operations are used to properly size the spillway and outlet works. The reservoir capacity is a major factor in flood routings and may determine the size and crest elevation of the spillway. The reservoir operation and reservoir capacity allocations will determine the location and size of outlet works for the controlled release of water for downstream requirements and flood control.

b. Area-capacity tables. Reservoir area-capacity tables should be prepared before the final designs and specifications are completed. These area-capacity tables should be based on the best available topographic data and should be the final design for administrative purposes until superseded by a reservoir resurvey. To ensure uniform reporting of data for design and construction, standard designations of water surface elevations and reservoir capacity allocations should be used.

14-4. Routing the Spillway Design Flood

a. Discharge facilities. The facilities available for discharging inflow from the spillway design flood depend on the type and design of the dam and its proposed use. A single dam installation may have two or more of the following discharge facilities: uncontrolled overflow spillway, gated overflow spillway, regulating outlet, and power plant. With a reservoir full to the spillway crest at the beginning of the design flood, uncontrolled discharge will begin at once. Surcharge storage is created when the outflow capacity is less than the inflow and the excess water goes into storage, causing the pool level to rise above the

spillway crest. The peak outflow will occur at maximum pool elevation, which should always be less, to some degree, than the peak inflow.

b. Gated spillway. With a gated spillway, the normal operating level is usually near the top of the gates, although at times it may be drawn below this level by other outlets. A gated spillway's main purpose is to maximize available storage and head, while at the same time limiting backwater damages by providing a high initial discharge capacity. In routing the spillway design flood, an initial reservoir elevation at the normal full pool operating level is assumed. Operating rules for spillway gates must be based on careful study to avoid releasing discharges that would be greater than would occur under natural conditions before construction of the reservoir. By gate operation, releases can be reduced and additional water will be held in storage, which is called "induced-surcharge storage." The release rates should be made in accordance with spillway gate regulation schedules developed for each gated reservoir. EM 1110-2-3600 Section 4-5 describes induced surcharge storage and the development and testing of the regulation schedules.

c. Surcharge storage. The important factor in the routing procedure is the evaluation of the effect of storage in the upper levels of the reservoir, surcharge storage, on the required outflow capacity. In computing the available storage, the water surface is generally considered to be level. There will be a sloping water surface at the head of the reservoir due to backwater effect, and this condition will create an additional "wedge storage." However, in most large and deep reservoirs this incremental storage can be neglected.

d. Drawdown. If a reservoir is drawn down at the time of occurrence of the spillway design flood, the initial increments of inflow will be stored with the corresponding reduction in ultimate peak outflow. Therefore, for maximum safety in design it is generally assumed that a reservoir will be full to the top of flood-control pool at the beginning of the spillway design flood.

e. Large flood-control storage reservations. There may be exceptions to the above criteria in the case of reservoirs with large reservations for flood-control storage. However, even in such cases, a substantial part (> 50 percent) of the flood-control storage should be considered as filled by runoff from antecedent floods. The effect on the economics and safety of the project should be analyzed before adopting such assumptions. ER 1110-8-2(FR) contains guidance on inflow design flood development and application.

f. Release rates. Assuming a reservoir can be significantly drawn down in advance of the spillway design flood by using a short-term flood warning system is generally not acceptable for several reasons. The volume that can be released is the product of the total rate of discharge at the dam times the warning time. Because the warning time is usually short, except on large rivers, the release rate must be the greatest possible without flood damage downstream. Even under the most favorable conditions, it is unlikely that the released volume will be significant, relative to the volume of the spillway design flood.

14-5. Sizing the Spillway

a. Storage and spillway capacity. In determining the best combination of storage and spillway capacity to accommodate the selected inflow design flood, all pertinent factors of hydrology, hydraulics, design, cost, and damage should be considered. In this connection and when applicable, consideration should be given to the following factors:

(1) The characteristics of the flood hydrograph.

(2) The damage which would result if such a flood occurred without the dam.

(3) The damage which would result if such a flood occurred with the dam in-place.

(4) The damage which would occur if the dam or spillway were breached.

(5) Effects of various dam and spillway combinations on the probable increase or decrease of damages above or below the dam.

(6) Relative costs of increasing the capacity of the spillways.

(7) The use of combined outlet facilities to serve more than one function.

b. Outflow characteristics. The outflow characteristics of a spillway depend on the particular device selected to control the discharge. These control facilities may take the form of an overflow weir, an orifice, a tube, or a pipe. Such devices can be unregulated, or they can be equipped with gates or valves to regulate the outflow.

c. Flood routing. After a spillway control of certain dimensions has been selected, the maximum spillway discharge and the maximum reservoir water level can be determined by flood routing. Other components of the spillway can then be proportioned to conform to the required capacity and the specific site conditions, and a complete layout of the spillway can be established. Cost estimates of the spillway and dam can then be made. Estimates of various combinations of spillway capacity and dam height for an assumed spillway type, and of alternative types of spillways, will provide a basis for the selection of the most economical spillway type and the optimum relation of spillway capacity to the height of the dam.

d. Maximum reservoir level. The maximum reservoir level can be determined by routing the spillway design flood hydrograph using sequential routing procedures and the proposed operation procedures. This is a basic step in the selection of the elevation of the crest of the dam, the size of the spillway, or both.

e. Peak rate of inflow. Where no flood storage is provided, the spillway must be sufficiently large to pass the peak of the flood. The peak rate of inflow is then of primary interest, and the total volume in the flood becomes less important. However, where a relatively large storage capacity above normal reservoir level can be made economically available by a higher dam, a portion of the flood volume can be retained temporarily in reservoir surcharge space, and the spillway capacity can be reduced considerably. If a dam could be made sufficiently high to provide storage space to impound the entire volume of the flood above normal storage level, theoretically, no spillway other than an emergency type would be required, provided the outlet capacity could evacuate the surcharge storage in a reasonable period of time in anticipation of a recurring flood. The maximum reservoir level would then depend entirely on the volume of the flood, and the rate of inflow would be of no concern. From a practical standpoint, however, relatively few sites will permit complete storage of the inflow design flood by surcharge storage.

f. Overall cost. The spillway length and corresponding capacity may have an important effect on the overall cost of a project because the selection of the spillway characteristics is based on an economic analysis. In many reservoir projects, economic considerations will necessitate a design utilizing surcharge. The most economical combination of surcharge storage and spillway capacity requires flood routing studies and economic studies of the costs of spillway-dam combinations. Among the many economic factors that may be considered are damage due to backwater in the reservoir, cost-height relations for gates, and utilization in the dam of material excavated from the spillway channel. However, consideration must still be given to the minimum size spillway which must be provided for safety.

g. Comprehensive study. The study may require many flood routings, spillway layouts, and spillway and dam estimates. Even then, the study is not necessarily complete because many other spillway arrangements could be considered. A comprehensive study to determine alternative optimum combinations and minimum costs may not be warranted for the design of some dams. Judgment on the part of the designer would be required to select for study only those combinations which show definite advantages, either in cost or adaptability. For example, although a gated spillway might be slightly cheaper than an ungated spillway, it may be desirable to adopt the latter because of its less complicated construction, its automatic and trouble-free operation, its ability to function without an attendant, and its less costly maintenance.

14-6. Outlet Works

a. Definition. An outlet works consists of the equipment and structures which together release the required water for a given purpose or combination of purposes. Flows through river outlets and canal or pipeline outlets change throughout the year and may involve a wide range of discharges under varying heads. The accuracy and ease of control are major considerations and a great amount of planning may be justified in determining the type of control devices that can be best utilized.

b. Description. Usually, the outlet works consist of an intake structure, a conduit or series of conduits through the dam, discharge flow control devices, and an energy dissipating device where required downstream of the dam. The intake structure includes a trash-rack, an entrance transition, and stop-logs or an emergency gate. The control device can be placed at the intake on the upstream face, at some point along the conduit and be regulated from galleries inside the dam, or at the downstream end of the conduit with the operating controls placed in a gate-house on the downstream face of the dam. When there is a power plant or other structure near the face of the dam, the outlet conduits can be extended farther downstream to discharge into the river channel beyond these features. In this case, a control valve may be placed in a gate structure at the end of the conduit.

c. Discharge. Discharges from a reservoir outlet works fluctuate throughout the year depending upon downstream water needs and reservoir flood control requirements. Therefore, impounded water must be released at specific regulated rates. Operating gates and regulating valves are used to control and regulate the outlet works flow and are designed to operate in any position from closed to fully open. Guard or emergency gates are designed to close if the operating gates fail, or where dewatering is desired to inspect or repair the operating gates.

d. Continuous low-flow releases. Continuous low-flow releases are usually required to satisfy the needs of fish, wildlife and existing water rights downstream from the dam. When the low-flow release is small, one or two separate small bypass pipes, with high-pressure regulating valves, are provided to facilitate operations. Flood-regulating gates may be used for making low-flow releases when those low-flow releases require substantial gate openings (EM 1110-2-1602).

e. Uses of an outlet works. An outlet works may be used for diverting the river flow or portion thereof during a phase of the construction period, thus avoiding the necessity for supplementary installations for that purpose. The outlet structure size dictated by this use rather than the size indicated for ordinary outlet requirements may determine the final outlet works capacity.

f. Intake level. The establishment of the intake level is influenced by several considerations such as maintaining the required discharge at the minimum reservoir operating elevation, establishing a silt retention space, and allowing selective withdrawal to achieve suitable water temperature and/or quality. Dams which will impound waters for irrigation, domestic use, or other conservation purposes must have the outlet works intake low enough to be able to draw the water down to the bottom of the allocated storage space. Further, if the outlets are to be used to evacuate the reservoir for inspection or repair of the dam, they should be placed as low as practicable. However, it is usual practice to make an allowance in a reservoir for inactive storage for silt deposition, fish and wildlife conservation, and recreation.

g. Elevation of outlet intake. Reservoirs become thermally stratified, and taste and odor vary between elevations. Therefore, the outlet intake should be established at the best elevation to achieve satisfactory water quality for the purpose intended. Downstream fish and wildlife requirements may determine the temperature at which the outlet releases should be made. Municipal and industrial water use increases the emphasis on water quality and requires the water to be drawn from the reservoir at the elevation which produces the most satisfactory combination of odor, taste, and temperature. Water supply releases can be made through separate outlet works at different elevations if requirements for the individual water uses are not the same and the reservoir is stratified.

h. Energy-dissipating devices. The two types of energy dissipating devices most commonly used in conjunction with outlet works on concrete dams are hydraulic jump stilling basins and plunge pools. On some dams, it is possible to arrange the outlet works in conjunction with the spillway to utilize the spillway-stilling device for dissipating the energy of the water discharging from the river outlets. Energy-dissipating devices for free-flow conduit outlet works are essentially the same as those for spillways.

Chapter 15
Dam Freeboard Requirements

15-1. Basic Considerations

a. Freeboard. Freeboard protects dams and embankments from overflow caused by wind-induced tides and waves. It is defined as the vertical distance between the crest of a dam and some specified pool level, usually the normal operating level or the maximum flood level. Depending on the importance of the structure, the amount of freeboard will vary in order to maintain structural integrity and the estimated cost of repairing damages resulting from overtopping. Riprap or other types of slope protection are provided within the freeboard to control erosion that may occur even without overtopping.

b. Estimating freeboard. Freeboard is generally based on maximum probable wind conditions when the reference elevation is the normal operating level. When estimating the freeboard to be used with the probable maximum reservoir level, a lesser wind condition is used because it is improbable that maximum wind conditions will occur simultaneously with the maximum flood level. A first step in wave height determinations is a study of available wind records to determine velocities and related durations and directions. Three basic considerations are generally used in establishing freeboard allowance. These are wave characteristics, wind setup, and wave runup.

c. Further information. The Corps of Engineers Coastal Engineering Research Center (CERC) has developed criteria and procedures for evaluating each of the above areas. The primary references are EM 1110-2-1412 and EM 1110-2-1414. The procedures presented in these manuals have received general acceptance for use in estimating freeboard requirements for reservoirs.

d. Applications. In applications for inland reservoirs, it is necessary to give special consideration to the influences that reservoir surface configuration, surrounding topography, and ground roughness may have on wind velocities and directions over the water surface. The effects of shoreline irregularities on wave refraction and influences of water depth variations on wave heights and lengths must be accounted for. Although allowances can only be approximated, the estimates of wave and wind tide characteristics in inland reservoirs can be prepared sufficiently accurate for engineering purposes.

15-2. Wind Characteristics over Reservoirs

a. General. The more violent windstorms experienced in the United States are associated with tropical storms (hurricanes) and tornadoes. Hurricane wind characteristics may affect reservoir projects located near Atlantic and Gulf coastlines, but winds associated with tornadoes are not applicable to the determination of freeboard allowances for wave action. In mountainous regions, the flow of air is influenced by topography as well as meteorological factors. These "orographic" wind effects, when augmented by critical meteorological patterns, may produce high wind velocities for relatively long periods of time. Therefore, they should be given special consideration in estimating wave action in reservoirs located in mountainous regions. In areas not affected by major topographic influences, air movement is generally the result of horizontal differences in pressure which in turn are due primarily to large-scale temperature differences in air masses. Wind velocities and durations associated with these meteorological conditions, with or without major influences of local topography, are of major importance in estimating wave characteristics in reservoirs.

b. Isovel patterns. Estimates of wind velocities and directions near a water surface at successive intervals of time, as a windstorm passes the area, may be established by deriving "isovel" patterns. Sequence relations can represent wind velocities at, say, one-half hour intervals during periods of maximum winds, and one-hour or longer intervals thereafter. The "isovel" lines connect points of equal wind speeds, resembling elevation contour maps. Wind directions are indicated by arrows. EM 1110-2-1412, Sections 1.9 and 1.10 describe storms and the storm surge generation process. Figure 1-1a shows an example wind isovel pattern and pertinent parameters.

c. Relation of wind duration to wave heights. If wind velocity over a particular fetch remains constant, wave heights will progressively increase until a limiting maximum value is attained, corresponding approximately to relations dependent on fetch distance, wind velocity, and duration. Accordingly, wind velocity-duration relations applicable to effective reservoir fetch areas are needed for use in computing wave characteristics in reservoirs.

d. Wind velocity-duration relations. In some cases it is desired to estimate wave characteristics in existing reservoirs in order to analyze causes of riprap damage or for other reasons. Wind records, supplemented by meteorological studies are usually required. Data on actual windstorms of record have been maintained at many U.S. Weather Bureau stations. Index values, such as the fastest mile, 1-min average or 5-min average velocities, with direction indications, are usually presented in climatological data publications. Some data collected by other agencies and private observers may be available in published or unpublished form. However, information regarding wind velocities sustained for several hours or days is not ordinarily published in detail. Accordingly, special studies are usually required to determine wind velocity-duration relations applicable to specific effective fetch areas involved in wave computations. Basic records for such studies are usually available from the U.S. Weather Bureau offices or other observer stations. Some summaries of wind velocities over relatively long periods of time have been published by various investigators, and others may be available in project reports related to water resources development.

e. Generalized wind velocity-duration relations. Studies show that maximum wind velocities in one general direction during major windstorms, in most regions of the United States, have averaged approximately 40 to 50 mph for a period of 1 hr. Corresponding velocities in the same general direction for periods of 2 hr and 6 hr have averaged 95 percent and 88 percent, respectively, of the maximum 1-hr average velocity. In EM 1110-2-1414, Figure 5-26 provides the ratio of wind speed of any duration to the 1-hr wind speed. Extreme wind velocities for brief periods, normally referred to as "fastest mile" or 1-min average, have been recorded as high as 150 to 200 percent of maximum 1-hr averages in most regions. In EM 1110-2-1414, Figures 5-18 through 5-20 provide the annual extreme fastest-mile speed 30 ft (9.1 m) above ground for the 25-year, 50-year, and 100-year recurrence intervals, respectively. However, these extreme values are seldom of interest in computing wave characteristics in reservoirs. Generalized wind velocity-duration relations are considered to be fairly representative of maximum values that are likely to prevail over a reservoir in generally a single direction for periods up to 6 hr (excluding projects located in regions that are subject to severe hurricanes or orographic wind-flow effects). Special studies of wind characteristics associated with individual project areas should be made when determinations of unusual importance, or problems requiring consideration of wind durations exceeding 6 hr, are involved.

f. Ratio of wind velocities over water and land areas. The wind velocities described in paragraph *d* are for over land. Under comparable meteorological conditions, wind velocities over water are higher than over land surfaces because of smoother and more uniform surface conditions. Winds blowing from land tend to increase with passage over reservoir areas, and vice versa. The relationships are not constant, but vary with topographic and vegetative cover of land areas involved, reservoir configurations, and other conditions affecting air flow. However, on the basis of research and field studies (Technical Memorandum No. 132, USACE 1962), the following ratios represent averages that are usually suitable for computing wave characteristics in reservoirs that are surrounded by terrain of moderate irregularities and surface roughness:

Fetch (F_e) in Miles	Wind ratio $\dfrac{\text{Over Water}}{\text{Over Land}}$
0.5	1.08
1	1.13
2	1.21
3	1.26
4	1.28
5 (or over)	1.30

15-3. Computation of Wave and Wind Tide Characteristics

a. Effective wind fetch (F_e) for wave generation. The characteristics of wind-generated waves are influenced by the distance that wind moves over the water surface in the "fetch" direction. The generally narrow irregular shoreline of inland reservoirs will have lower waves than an open coast because there is less water surface for the wind to act on. The method to compensate for the reduced water surface for an enclosed body of water is computation of an effective fetch. The effective fetch (F_e) adjusts radial lines from the embankment to various points on the reservoir shore. The radials spanning 45 deg on each side of the central radial are adjusted by the cosign of their angle to the central radial to estimate an average effective fetch. The computation procedure is shown in EM 1110-2-1414, Figure 5-33 and Example Problem 7-2. Generalized relations are based on effective fetch distances derived in this manner.

b. Fetch distance for wind tide computations. Fetch distances for use in estimating wind tide (set-up) effects are usually longer than effective fetch distances used in estimating wave heights. In as much as wind tide effects in deep inland reservoirs are relatively small, extensive studies to refine estimates are seldom justified. For practical purposes, it is usually satisfactory to assume that the wind tide fetch is equal to twice the effective fetch (F_e). If wind tide heights determined in this manner are relatively large in relation to overall freeboard requirements, more detailed analyses are advisable using methods as generally discussed in Chapter 3 of EM 1110-2-1414.

c. Generalized diagrams for wave height and wave period in deep inland reservoirs. EM 1110-2-1414, Figure 5-34, presents generalized relations between significant wave height, wave period, fetch (F_e), and wind velocities corresponding to critical durations. These diagrams were developed from research and field studies based on wind speed at 10 m (33 ft). If wind speeds are for a different level, Equation 5-12 can be used to adjust to the 10 m level (EM 1110-2-1414).

d. Wave characteristics in shallow inland lakes and reservoirs. In the analyses of wave characteristics, lakes and reservoirs are considered to be shallow when depths in the wave-generating area are generally less than about one-half the theoretical deep-water wave length (L_o) corresponding to the same wave period (T). Curves presented in Figures 5-35 through 5-44 represent relations between wave characteristics, fetch distances (in feet) and constant water depths in the wave-generating area, ranging from 5 to 50 ft (1.5 to 15.2 m) (EM 1110-2-1414).

e. Wind tides (set-up) in inland waters.

(1) When wind blows over a water surface, it exerts a horizontal stress on the water, driving it in the direction of the wind. In an enclosed body of water, this wind effect results in a piling up of water at the leeward end, and a lowering of water level at the windward end. This effect is called "wind tide" or "wind set-up." Wind set-up can be reasonably estimated for lakes and reservoirs, based on the following equation:

$$S = \frac{U^2 \, F}{1400 \, D} \qquad (15\text{-}1)$$

in which S is wind tide (set-up) in feet above the stillwater level that would prevail with zero wind action; U is the average wind velocity in statute miles per hour over the fetch distance (F) that influences wind tide; D is the average depth of water generally along the fetch line (EM 1110-2-1414). The fetch distance (F) used in the above formula is usually somewhat longer than the effective fetch (F_e) used in wave computations, as indicated in paragraph 15-3b. Refer to EM 1110-2-1414, Section 3-2 for a discussion of prediction models.

15-4. Wave Runup on Sloping Embankment

a. Introduction. Most dam embankments are fronted by deep water, have slopes between 1 on 2 and 1 on 4, and are armored with riprap. Rock-fill dams are considered as permeable rubble slopes and earth-fill dams with riprap armor are considered impermeable. Laboratory tests of many slopes, wave conditions, and embankment porosity provide sufficient data to make estimates of wave runup on a prototype embankment.

b. Relative runup relations. EM 1110-2-2904 Plate 25 presents generalized relations on wave runup on rubble-mound breakwaters and smooth impervious slopes. Plate 26 provides similar curves for various embankment slopes for water depths greater than three-times wave height. The curves correspond to statistical averages of a large number of small and large-scale hydraulic model test results, and have been adjusted for model scale effects to represent prototype conditions. The relations were based primarily on tests involving mechanically generated waves and may differ somewhat from relations associated with individual waves in natural wind-generated spectrums of waves. However, general field observations and comparisons with wave experiences support the conclusion that relations presented.

c. Runup of waves on sloping embankments.

(1) If waves generated in deep water (i.e., depths exceeding about one-third to one-half the wave length), reach the toe of an embankment without breaking, the vertical height of runup may be computed by multiplying the deep-water wave height (H_o) by the relative runup ration (R/H) obtained from EM 1110-2-2904, Plate 26, for the appropriate slope and wave steepness (H_o/L_o). In this case, deep-water values of H_o and L_o should be used as indices, even though wave heights and lengths are modified by passing through areas in which water depths are less than $L_o/3$ (provided the depth is not small enough to cause the wave to break before reaching the embankment). That is, the height of runup may be computed by using the deep-water steepness H_o/L_o, whether the structure under study is located in deep water or in shallow water, provided the wave does not break before reaching the toe of the structure.

(2) If waves are generated primarily by winds over open-water areas where the relative depth (d/L) is appreciably less than 0.3, the wave heights and periods should be computed by procedures applicable to shallow waters.

(3) Waves generated by wind over open-water areas of a particular depth change characteristics when they reach areas where the constant depth is substantially less, the height (H) tending to increase while the length (L) decreases. The distribution of wave energy changes as a wave enters the shallow water, the proportion of total energy which is transmitted forward with the wave toward the shore increasing, although the actual amount of this

translated energy remains constant except from minor frictional effects. If the depth continues to decrease, the steepness ratio (H/L) increases, until finally the wave becomes unstable and breaks, resulting in appreciable energy dissipation. Theoretically, the maximum wave cannot exceed 0.78 D, where D is the depth of water without wave action. After breaking, the waves will tend to reform with lower heights within a distance equal to a few wave lengths. For most engineering applications, it is satisfactory to assume that the wave height after breaking will equal approximately 0.78 D in the shallow area and that L will be the same as before the wave broke. Plate 26 would then be entered with a wave-steepness ratio equal to 0.78/L to determine the relative runup ration (R/H), and this ratio would be multiplied by 0.78 D to obtain the estimated runup height (R). This procedure should provide conservative results under circumstances in which the distance between the point where waves reach breaking depths and location of the structure under study is long enough to permit waves to reform, and short enough to preclude substantial build-up by winds prevailing over the shallower area. More accurate values could be obtained by using this breaking height (0.78 D) and period to obtain comparable deep water values of H_o and L_o.

d. Adjustments in wave runup estimates for variations in riprap.

(1) A rough riprap layer on an embankment tends to reduce the height of runup after a wave breaks. If the riprap layer thickness is small in comparison with wave magnitudes and the underlying surface is relatively impermeable, so that the void spaces in the riprap remain mostly filled with water between successive waves during severe storm events, the height of runup may closely approach heights attained on smooth embankments of comparable slope. However, if the riprap layer is sufficiently rough, thick, and free draining to quickly absorb the water that impinges on the embankment as each successive wave breaks, further wave runup will be almost completely eliminated.

(2) The design of riprap to absorb most of the energy of breaking waves is practicable if waves involved will be relatively small or moderate, but costs and other practical considerations usually preclude such design where large waves are encountered. Accordingly, the design characteristics of riprap layers are usually somewhere between the two extremes described above.

15-5. Freeboard Allowances for Wave Action

a. Purpose. In connection with the design of dams and reservoirs, the estimate of freeboard is required to establish allowances needed to provide for wave action that is likely to affect various project elements, as follows:

(1) Main embankment of the dam, and supplemental dike sections.

(2) Levees that protect areas within potential flowage limits of the reservoir.

(3) Highway and railroad embankments that intersect the reservoir limits.

(4) Structures located within the reservoir area.

(5) Shoreline areas that are subject to adverse effects of wave action.

b. Freeboard on dams. The establishment of freeboard allowances on dams includes not only the consideration of potential wave characteristics in a reservoir, but several other factors of importance, including certain policy matters.

c. Freeboard allowances for wave action on embankments and structures within reservoir flowage limits.

(1) Wave action effects must be taken into account in establishing design grades and slope protection measures for highway, railroad, levee, and other embankments that intersect or boarder a reservoir. The design of operating structure, boat docks, recreational beaches, and shoreline protection measures at critical locations involves the consideration of wave characteristics and frequencies under a range of conditions. Estimates of wave characteristics affecting the design of these facilities can have a major influence on the adequacy of design and costs of relocations required for reservoir projects, and in the development of supplemental facilities.

(2) The freeboard reference level selected as a base for estimating wave effects associated with each of the several types of facilities referred to above will be governed by considerations associated with the particular facility. Otherwise, procedures generally as described with respect to the determination of freeboard allowances for dams should be followed, and stage hydrographs and related wave runup elevations corresponding to the selected wind criteria should be prepared. However, the freeboard reference level and coincident wind velocity-duration relations selected for these studies usually correspond to conditions that would be expected with moderate frequency, instead of the rare combinations assumed in estimating the height of dam required for safety.

(3) In estimating effects of wave action on embankments and structures, the influences of water depths near the facility should be carefully considered. If the shallow depths prevail for substantial distances from the embankment or structure under study, wave effects may be greatly reduced from those prevailing in deep-water areas. On the other hand, facilities located where sudden reductions in water depths cause waves to break are likely to be subjected to greater dynamic forces than would be imposed on similar facilities located in deep water. This consideration is particularly important in estimating the effects waves may have on bridge structures that are partially submerged under certain reservoir conditions.

(4) Systematic analyses of wave effects associated with various key locations along embankments that cross or border reservoirs provide a practical basis for varying design grades and erosion protection measures to establish the most economical plan to meet pertinent operational and maintenance standards.

Chapter 16
Dam Break Analysis

16-1. Introduction

a. *Corps policy.* It is the policy of the Corps of Engineers to design, construct, and operate dams safely (ER 1110-8-2(FR)). When a dam is breached, catastrophic flash flooding occurs as the impounded water escapes through the gap into the downstream channel. Usually, the response time available for warning is much shorter than that for precipitation-runoff floods, so the potential for loss of life and property damage is much greater.

b. *Hazard evaluation.* A hazard evaluation is the basis for selecting the performance standards to be used in dam design or in evaluating existing dams. When flooding could cause significant hazards to life or major property damage, the design flood selected should have virtually no chance of being exceeded. ER 1110-8-2(FR) provides dam safety standards with respect to the appropriate selection of an inflow design flood. If human life is at risk, the general requirement is to compute the flood using PMP. If lesser hazards are involved, a smaller flood may be selected for design. However, all dams should be designed to withstand a relatively large flood without failure even when there is apparently no downstream hazard involved under present conditions of development.

c. *Safety design.* Safety design includes studies to ascertain areas that would be flooded during the design flood and in the event of dam failure. The areas downstream from the project should be evaluated to determine the need for land acquisition, flood plain management, or other methods to prevent major damage. Information should be developed and documented suitable for releasing to downstream interests regarding the remaining risks of flooding.

d. *National Dam Safety Act.* The potential for catastrophic flooding due to dam failures in the 1960's and 1970's brought about passage of the National Dam Safety Act, Public Law 92-367. The Corps of Engineers became responsible for inspecting U.S. Federal and non-Federal dams, which met the size and storage limitations of the act, in order to evaluate their safety. The Corps inventoried dams; surveyed each State and Federal agency's capabilities, practices, and regulations regarding the design, construction, operation, and maintenance of the dams; developed guidelines for the inspection and evaluation of dam safety; and formulated recommendations for a comprehensive national program.

e. *Flood emergency documents.* In support of the National Dam Safety Program, flood emergency planning for dams was evaluated in the 1980's, and a series of documents were published: *Emergency Planning for Dams, Bibliography and Abstracts of Selected Publications* (HEC 1982a), *Flood Emergency Plan, Guidelines for Corps Dams*, Research Document 13 (HEC 1980), *Example Emergency Plan for Blue Marsh Dam and Lake* (HEC 1983a), and *Example Plan for Evacuation of Reading, Pennsylvania in the Event of Emergencies at Blue Marsh Dam and Lake* (HEC 1983b). The development of an emergency plan requires the identification of the type of emergencies to be considered, the gathering of needed data, performing the analyses and evaluations, and presenting the results. HEC Research Document 13 (HEC 1980) provides guidelines for each step of the process.

16-2. Dam Breach Analysis

a. *Causes of dam failures.* Dam failures can be caused by overtopping a dam due to insufficient spillway capacity during large inflows to the reservoir, by seepage or piping through the dam or along internal conduits, slope embankment slides, earthquake damage and liquification of earthen dams from earthquakes, or landslide-generated waves within the reservoir. Hydraulics, hydrodynamics, hydrology, sediment transport mechanics, and geotechnical aspects are all involved in breach formation and eventual dam failure. HEC Research Document 13 lists the prominent causes as follows:

(1) Earthquake.

(2) Landslide.

(3) Extreme storm.

(4) Piping.

(5) Equipment malfunction.

(6) Structural damage.

(7) Foundation failure.

(8) Sabotage.

b. *Dam breach characteristics.* The breach is the opening formed in the dam when it fails. Despite the fact that the main modes of failure have been identified as piping or overtopping, the actual failure mechanics are not well understood for either earthen or concrete dams. In previous attempts to predict downstream flooding due to dam failures, it was usually assumed that the dam failed completely and instantaneously. These assumptions of instantaneous and complete breaches were used for reasons of convenience when applying certain mathematical techniques for analyzing dam-break flood waves. The presumptions are somewhat appropriate for concrete arch-type dams, but they are not suitable for earthen dams and concrete gravity-type dams.

(1) Earthen dams, which exceedingly outnumber all other types of dams, do not tend to completely fail, nor do they fail instantaneously. Once a developing breach has been initiated, the discharging water will erode the breach until either the reservoir water is depleted or the breach resists further erosion. The fully formed breach in earthen dams tends to have an average width (b) in the range ($h_d < b < 3h_d$) where, h_d is the height of the dam. Breach widths for earthen dams are therefore usually much less than the total length of the dam as measured across the valley. Also, the breach requires a finite interval of time for its formation through erosion of the dam materials by the escaping water. The total time of failure may range from a few minutes to a few hours, depending on the height of the dam, the type of materials used in construction, and the magnitude and duration of the flow of escaping water. Piping failures occur when initial breach formation takes place at some point below the top of the dam due to erosion of an internal channel through the dam by escaping water. As the erosion proceeds, a larger and larger opening is formed. This is eventually hastened by caving-in of the top portion of the dam.

(2) Concrete gravity dams also tend to have a partial breach as one or more monolith sections formed during the dam construction are forced apart by the escaping water. The time for breach formation is in the range of a few minutes.

(3) Poorly constructed earthen dams and coal-waste slag piles which impound water tend to fail within a few minutes and have average breach widths in the upper range or even greater than those for the earthen dams mentioned above.

c. Dam breach parameters. The parameters of failure depend on the dam and the mode of failure. For flood hydrograph estimation, the breach is modeled assuming weir conditions, and the breach size, shape, and timing are the important parameters. The larger the breach opening and the shorter the time to total failure, the larger the peak outflow. HEC Research Document 13, Table 1, lists suggested breach parameters for earth-fill, concrete-gravity, and concrete-arch dams. There are two basic approaches used to determine possible breach sizes and times.

(1) The first approach uses statistically derived regression equations, like those formulated by MacDonald and Langridge-Monopolis (1984) and by Froelich (1987). Both sets of equations are based on actual data from dozens of historic dam failures. The MacDonald, and Langridge-Monopolis study was based on data from 42 constructed earth- and rock-fill dams. The Froelich study included data from constructed and landslide-formed earthen dams. Both studies resulted in a set of graphs and equations that can be used to predict the approximate size of the breach and the time it takes for the breach to reach its full size.

(2) The second approach is a physically based computer model called BREACH, developed by Dr. Danny Fread (1989) for the National Weather Service. The breach model uses sediment transport and hydraulic routing equations to simulate the formation of either a piping or overtopping type of failure. The model requires information about the physical dimensions of the dam, as well as a detailed description of the soil properties of the dam. Soils information includes D50 (mm), porosity, unit weight (lb/ft³), internal friction angle, cohesive strength (lb/ft²), and D90/D30. These parameters can be specified separately for the inner-core and outside-bank materials of a dam.

16-3. Dam Failure Hydrograph

a. Flow hydrograph. The flow hydrograph from a breached dam may be computed using traditional methods for flow routing through a reservoir and downstream channel. The reservoir routing approach is the same as routing for the spillway design flood, described in Chapter 14. Generally, a short time step is required because the breach formation and resulting reservoir outflow change rapidly with time.

b. Routing methods. The choice between hydraulic and hydrologic routing depends on many factors, including the nature of available data and accuracy required. The hydraulic method is the more accurate method of routing the unsteady flow from a dam failure flood through the downstream river. This technique simultaneously computes the discharge, water surface elevation, and velocity throughout the river reach. Chapter 9 of EM 1110-2-1417 describes the routing methods and applicability of routing techniques. Chapter 5 of EM 1110-2-1416 describes unsteady flow computations.

c. Geometry and surface area. The geometry and surface area of the reservoir can also affect the choice of method. For very narrow and long reservoirs where the dam is relatively large, the change of water level at the failed dam is rapid, and the unsteady flow method is useful. However, for very large reservoirs where the dam is small compared to the area of the lake, the change in water level is relatively slow and the storage routing method (Modified Puls) is economical in developing the failure hydrograph. Because of the rapid change in water level, small time periods are required for both methods.

d. Height of downstream water. The height of the water downstream of a dam (tailwater) also affects the outflow hydrograph in a failure analysis. It also affects the formation or nonformation of a bore in front of the wave.

e. Deriving the peak outflow. By assuming a rectangular cross section, zero bottom slope, and an instantaneous failure of a dam, the peak outflow can be derived by the mathematical expression originally developed by St. Venant, as follows:

$$Q_{max} = \frac{8}{27} W_b \sqrt{g} Y_o^{3/2} \qquad (16\text{-}1)$$

where

Y_o is the initial depth, W_b is the width of the breach, g is the gravity coefficient, and the water depth, y, just downstream of the dam is

$$y = \frac{4}{9} Y_o \qquad (16\text{-}2)$$

This equation is applicable only for relatively long and narrow rectangular channels where the dam is completely removed. *Guidelines for Calculating and routing a Dam-Break Flood*, HEC Research Document No. 5 (HEC 1977) describes this approach.

f. Failed dam outflow hydrograph. The outflow hydrograph from a failed dam may also be approximated by a triangle. For instantaneous failure, a right triangle is applicable. The base represents the time to empty the reservoir volume, and the height represents the instantaneous peak outflow. In erosion analysis, the Office of Emergency Services, after consultation with other agencies, suggested an isosceles triangle. The rising side of the isosceles triangle is developed by assuming that half of the reservoir storage is required to erode the dam to natural ground level. The apex of the triangle represents the peak flow through the breach under the assumption that the flow occurs at critical depth.

g. Potential for overtopping. The Hydrologic Engineering Center's HEC-1 Flood Hydrograph Package (HEC 1990c) can be used to determine the potential for overtopping of dams by run off resulting from various proportions of the PMF. This technique is most appropriate for simulating breaches in earthen dams caused by overtopping. Other conditions may be approximated, however, such as instantaneous failure. This method makes six assumptions:

(1) Level-pool reservoir routing to determine time-history of pool elevation.

(2) Breach shape is a generalized trapezoid with bottom width and side slopes prespecified by the analyst.

(3) Bottom of the breach moves downward at a constant rate.

(4) Breach formation begins where the water surface in the reservoir reaches a prespecified elevation.

(5) Breach is fully developed when the bottom reaches a prespecified elevation.

(6) Discharge through the breach can be calculated independently of downstream hydraulics, i.e., critical depth occurs at or near the breach. A tailwater rating curve or a single cross section (assuming normal-depth for a rating) can be used to simulate submergence effects.

The total discharge from the dam at any instant is calculated by summing the individual flows through the low level outlet, over the spillway and top of the dam, and through the breach.

h. Peak flow values. With several calculations of theoretical flood peaks from assumed breaches, peak flow values may seem either too low or too high. One way of checking the reasonableness of the assumption is to compare the calculated values with historical failures. An envelope of estimated flood peaks from actual dam failures prepared by the Bureau of Reclamation is a good means of comparing such values. HEC Research Document No. 13, Figure 2, provides an envelope of experienced outflow rates from breached dams, as a function of hydraulic depth.

16-4. Dam Break Routing

a. Dam-break flood hydrographs. Dam-break flood hydrographs are dynamic, unsteady flow events.

Therefore, the preferred routing approach is to utilize a full unsteady flow routing model. The HEC-1 Flood Hydrograph package provides the capability to compute and route the inflow design flood and compute the breach and resulting hydrograph, but its channel routing is limited to hydrologic methods. The most appropriate HEC-1 approach is the Muskingum-Cunge option. The option uses a simple cross section plus reach slope and length to define a routing reach. No downstream backwater effects are considered. If simplified representations of the downstream river reaches are acceptable, an adequate routing may be obtained.

b. St. Venant equations. The St. Venant equations apply to gradually varied flow with a continuous profile. If features which control or interrupt the water surface profile exist along the main stem of the river or its tributaries, internal boundary conditions are required. These features include dams, bridges, roadway embankments, etc. If the structure is a dam, the total discharge is the sum of spillway flow, flow over the top of the dam, gated-spillway flow, flow through turbines, and flow through a breach, should a breach occur. The spillway flow and dam overtopping are treated as weir flow, with corrections for submergence. The gated outlet can represent a fixed gate or one in which the gate opening can vary with time. These flows can also be specified by rating curves which define discharge passing through the dam as a function of upstream water surface elevation.

c. Unsteady flow computer programs. There are an increasing number of available unsteady flow computer programs. The FLDWAV program is a generalized unsteady-flow simulation model for open channels. It replaces the DAMBRK, DWOPER, and NETWORK models, combining their capabilities and providing new hydraulic simulation procedures within a more user-friendly model structure (DeVries and Hromadka 1993). Given the long history of application by the National Weather Service, this program is likely the most capable for this purpose.

d. FLDWAV. FLDWAV can simulate the failure of dams caused by either overtopping or piping failure of the dam. The program can also represent the failure of two or more dams located sequentially on a river. The program is based on the complete equations for unsteady open-channel flow (St. Venant equations). Various types of external and internal boundary conditions are programmed into the model. At the upstream and downstream boundaries of the model (external boundaries), either discharges or water surface elevations, which vary with time, can be specified.

e. Special features. The following special features and capacities are included in FLDWAV: variable Δt and Δx computational intervals; irregular cross-sectional geometry; off-channel storage; roughness coefficients that vary with discharge or water surface elevation, and with distance along the waterway; capability to generate linearly interpolated cross sections and roughness coefficients between input cross sections; automatic computation of initial steady flow and water elevations at all cross sections along the waterway; external boundaries of discharge or water surface elevation time series (hydrographs), a single-valued or looped depth-discharge relation (tabular or computed); time-dependent lateral inflows (or outflows); internal boundaries enable treatment of time-dependent dam failures, spillway flows, gate controls, or bridge flows, or bridge-embankment overtopping flow; short-circuiting of floodplain flow in a valley with a meandering river; levee failure and/or overtopping; a special computational technique to provide numerical stability when treating flows that change from supercritical to subcritical , or conversely, with time and distance along the waterway; and an automatic calibration technique for determining the variable roughness coefficient by using observed hydrographs along the waterway.

f. UNET. The unsteady flow program UNET (HEC 1995) has a dam-break routing capability. However, there has been limited application of this feature. UNET could be used to route the outflow hydrograph computed in an HEC-1 runoff-dam break model. Both programs can read and write hydrographs using the HEC Data Storage System, HEC-DSS (HEC 1995a).

16-5. Inundation Mapping

a. Preparation of maps. To evaluate the effects of dam failure, maps should be prepared delineating the area which would be inundated in the event of failure. Land uses and significant development or improvements within the area of inundation should be indicated. The maps should be equivalent to or more detailed than the USGS quadrangle maps, 7.5-min series, or of sufficient scale and detail to identify clearly the area that should be evacuated if there is evident danger of failure of the dam. Copies of the maps should be distributed to local government officials for use in the development of an evacuation plan. The intent of the maps is to develop evacuation procedures in case of collapse of the dam, so the travel time of the flood wave should be indicated on every significant habitation area along the river channel.

b. Evaluation of hazard potential. To assist in the evaluation of hazard potential, areas delineated on inundation maps should be classified in accordance with the degree of occupancy and hazard potential. The potential for loss of life is affected by many factors, including but not limited to the capacity and number of exit roads to higher ground and available transportation. Hazard potential is greatest in urban areas. The evaluation of hazard potential should be conservative because the extent of inundation is usually difficult to delineate precisely.

c. Hazard potential for recreation areas. The hazard potential for affected recreation areas varies greatly, depending on the type of recreation offered, intensity of use, communications facilities, and available transportation. The potential for loss of life may be increased where recreationists are widely scattered over the area of potential inundation because they would be difficult to locate on short notice.

d. Industries and utilities. Many industries and utilities requiring substantial quantities of water are located on or near rivers or streams. Flooding of these areas and industries, in addition to causing the potential for loss of life, can damage machinery, manufactured products, raw materials and materials in process of manufacture, plus interrupt essential community services.

e. Least hazard potential. Rural areas usually have the least hazard potential. However, the potential for loss of life exists, and damage to large areas of intensely cultivated agricultural land can cause high economic loss.

f. Evacuation plans.

(1) Evacuation plans should be prepared and implemented by the local jurisdiction controlling inundation areas. The assistance of local civil defense personnel, if available, should be requested in preparation of the evacuation plan. State and local law enforcement agencies usually will be responsible for the execution of much of the plan and should be represented in the planning effort. State and local laws and ordinances may require that other state, county, and local government agencies have a role in the preparation, review, approval, or execution of the plan. Before finalization, a copy of the plan should be furnished to the dam agency or owner for information and comment.

(2) Evacuation plans will vary in complexity in accordance with the type and degree of occupancy in the potentially affected area. The plans may include delineation of the area to be evacuated; routes to be used; traffic control measures; shelter; methods of providing emergency transportation; special procedures for the evacuation and care of people from institutions such as hospitals, nursing homes, and prisons; procedures for securing the perimeter and for interior security of the area; procedures for the lifting of the evacuation order and reentry to the area; and details indicating which organizations are responsible for specific functions and for furnishing the materials, equipment, and personnel resources required. HEC Research Documents 19 and 20 provide example emergency plans and evacuation plans, respectively (HEC 1983a and b).

Chapter 17
Channel Capacity Studies

17-1. Introduction

a. General. Channel capacity studies tend to focus on high flows. Flood operations for a reservoir will require operational downstream targets for nondamaging flows when excess water must be released. Nondamaging channel capacity may be defined at several locations, and the target flow may be defined at several levels. There may be lower targets for small flood events and, under extreme flood situations, the nondamaging target may cause some minor damage. Also, the nondamaging flow target may vary seasonally and depend on floodplain land use.

b. Withstanding release rates. Channel capacity is also concerned with the capability of the channel to withstand reservoir release rates. Of particular concern is the reach immediately downstream from the reservoir. High release rates for hydropower or flood control could damage channel banks and cause local scour and channel degradation.

c. Channel capacity. While flood operation may focus on maximum channel capacity, planning studies usually require stage-discharge information over the entire range of expected operations. Also, low-flow targets may be concerned with maintaining minimum downstream flow depth for navigation, recreation, or environmental goals. Channel capacity studies typically provide information on safe channel capacity and stage-discharge (rating) curves for key locations.

17-2. Downstream Channel Capacity

a. Downstream channel erosion. Water flowing over a spillway or through a sluiceway is capable of causing severe erosion of the stream bed and banks below the dam. Consequently, the dam and its appurtenant works must be so designed that harmful erosion is minimized. The outlet works for a dam usually require an energy-dissipating structure. The design may vary from an elaborate multiple-basin arrangement to a simple head wall design, depending on the number of conduits involved, the erosion resistance of the exit channel bed material, and the duration, intensity, and frequency of outlet flows. A stilling basin may be provided for outlet works when such downstream uses as navigation, irrigation, and water supply, require frequent operation or when the channel immediately downstream is easily eroded. Sections 4-2b and 4-3j of EM 1110-2-3600 provide a general discussion of energy dissipators for spillways and outlet works, respectively.

b. Adequate capacity. The channel downstream should have adequate capacity to carry most flows from reservoir releases. After the water has lost most of its energy in the energy-dissipating devices, it is usually transported downstream through the natural channel to its destination points. With the expected release rates, the channel should be able to resist excessive erosion and scour, and have a large enough capacity to prevent downstream flooding except during large floods.

c. River surveys. River surveys of various types provide the basic physical information on which river engineering planning and design are based. Survey data include information on the horizontal configuration (planform) of streams; characteristics of the cross sections (channel and overbank); stream slope; bed and bank materials; water discharge; sediment characteristics and discharge; water quality; and natural and cultural resources.

d. Evaluating bank stability. To evaluate bank stability, it is essential to understand the complex historical pattern of channel migration and bank recession of the stream and the relationship of channel changes to streamflow. Studies of bank caving, based on survey data and aerial photographs, provide information on the progressively shifting alignment of a stream and are basic to laying out a rectified channel alignment. The concepts and evaluation procedures presented in "Stability of Flood Control Channels" (USACE 1990) are applicable to the channel capacity evaluation.

e. Interrupted sediment flow. A dam and reservoir project tends to interrupt the flow of sediment, which can have a significant impact on the downstream channel capacity. If the project is relatively new, the affect may not be seen by evaluating historic information or current channel conditions. The future channel capacity will depend on the long-term trends in aggradation and degradation along the river. General concepts on sediment analysis are presented in Chapter 9. *Sediment Investigations of Rivers and Reservoirs*, EM 1110-2-4000, is the primary reference for defining potential problems and analyses procedures.

f. Downstream floodplain land use. Channel capacity also depends on the long-term trends in downstream floodplain land use. While it is not a hydrologic problem, channel capacity studies should recognize the impact of floodplain encroachments on what is considered the nondamaging channel capacity. Anecdotal history has shown that many Corps' projects are not able to make planned channel-capacity releases due to development and encroachments downstream.

17-3. Stream Rating Curve

a. Stage-discharge relationship. The relationship between stage and discharge, the "rating" at a gauging station, is based on field measurements with a curve fitted to plotted data of stage versus discharge. For subcritical flow, the stage-discharge relationship is controlled by the stream reach downstream of the gauge; for supercritical flow, the control is upstream of the gauge. The stage-discharge relationship is closely tied to the rate of change of discharge with time, and the rating curve for a rising stage can be different from that for the falling stage in alluvial rivers.

b. Tailwater rating curve. The tailwater rating curve, which gives the stage-discharge relationship of the natural stream below the dam, is dependent on the natural conditions along the stream and ordinarily cannot be altered by the spillway design or by the release characteristics. Degradation or aggradation of the river below the dam, which will affect the ultimate stage-discharge conditions, must be recognized in selecting the tailwater rating curves to be used for design. Usually, river flows which approach the maximum design discharges have never occurred, and an estimate of the tailwater rating curve must either be extrapolated from known conditions or computed on the basis of assumed or empirical criteria. Thus, the tailwater rating curve at best is only approximate, and factors of safety to compensate for variations in tailwater must be included in dependent designs.

c. Extrapolation. Extrapolation of rating curves is necessary when a water level is recorded below the lowest or above the highest gauged level. Where the cross section is stable, a simple method is to extend the stage-area and stage-velocity curve and, for given stage values, take the product of velocity and cross-section area to give discharge values beyond the stage values that have been gauged. Generally, water-surface profiles should be computed to develop the rating beyond the range of observed data.

d. Rating curve shifts. The stage-discharge relationship can vary with time, in response to degradation, aggradation, or a change in channel shape at the control section, deposition of sediment causing increased approach velocities in a weir pond, vegetation growth, or ice accumulation. Shifts in rating curves are best detected from regular gauging and become evident when several gaugings deviate from the established curve. Sediment accumulation or vegetation growth at the control will cause deviations which increase with time, but a flood can flush away sediment and aquatic weed and cause a sudden reversal of the rating curve shift.

e. Flow magnitude and bed material. Stream bed configuration and roughness in alluvial channels are a function of the flow magnitude and bed material. Bed forms range from ripples and dunes in the lower regime (Froude number < 1.0) to a smooth plane bed, to antidunes with standing waves (bed and water surface waves in phase) and with breaking waves and, finally, to a series of alternative chutes and pools in the upper regime as the Froude number increases.

f. Upper and lower rating portions. The large changes in resistance to flow that occur as a result of changing bed roughness affect the stage-discharge relationship. The upper portion of the rating is relatively stable if it represents the upper regime (plane-bed, transition, standing wave, or antidune regime) of bed form. The lower portion of the rating is usually in the dune regime, and the stage-discharge relationship varies almost randomly with time. Continuous definition of the stage-discharge relationship at low flow is a very difficult problem, and a mean curve for the lower regime is frequently used for gauges with shifting control.

g. Break up of surface material. In gravel-bed rivers, a flood may break up the armoring of the surface gravel material, leading to general degradation until a new armoring layer becomes established and ratings tend to shift between states of quasi-equilibrium. It may then be possible to shift the rating curve up or down by the change in the mean-bed level, as indicated by plots of stage and bed level versus time.

h. Ice. Ice at the control section may also affect the normal stage-discharge relationship. Ice effects vary with the quantity and the type of ice (surface ice, frazil ice, or anchor ice). When ice forms a jam in the channel and submerges the control or collects in sufficient amounts between the control and the gauge to increase resistance to flow, the stage-discharge relationship is affected; however, ice may form so gradually that there is little indication of its initial effects. Surface ice is the most common form and affects station ratings more frequently than frazil ice or anchor ice. The major effect of ice on a rating curve is due to backwater and may vary from day to day.

17-4. Water Surface Profiles

a. Appropriate methods. For most channel-capacity studies, water surface profiles will be computed to develop the required information. Given the technical concerns described in the preceding section on rating curves, the selection of the appropriate method requires some evaluation of the physical system and the expected use of the information. The modeling methods are described in Chapter 8 and are presented in EM 1110-2-1416. While steady-flow water surface profiles are used in a majority of profile calculations, the unsteady flow aspects of reservoir operation or the long-term effects of changes in sediment transport may require the application of methods that capture those aspects.

b. Further information. The Corps, and other agencies, have accumulated considerable experience with river systems. Appendix D, "River Modeling - Lessons Learned" (EM 1110-2-1416), provides an overview of technical issues and modeling impacts that apply to profile calculations. *Stability of Flood Control Channels* (USACE 1990) provides case examples of stream stability problems, causes, and effects. While the focus is not on reservoirs, the experience reflects the high flow conditions that are a major concern with reservoir operation. And EM 1110-2-4000 provides procedures for problem assessment and modeling. All of these documents should be reviewed prior to formulating and performing technical studies.

Chapter 18
Real Estate and Right-of-Way Studies

18-1. Introduction

a. General. This chapter provides guidance on the application of hydrologic engineering principles to determine real estate acquisition requirements for reservoir projects. Topics include selection of the analysis method, potential problems, evaluation criteria, and references associated with the acquisition of real estate for reservoir projects developed by the Corps of Engineers.

b. Related documents. Real estate reporting requirements associated with feasibility reports, General Design Memoranda, and Real Estate Design Memoranda are set forth in ER 405-1-12. Real estate reporting requirements associated with the acquisition of lands downstream from spillways are set forth in ER 1110-2-1451, paragraph 9.

18-2. Definition of Terms

A list of terms and definitions used in this chapter is as follows:

a. Project design sediment. The volume and distribution of sediment deposited in a reservoir over the life of the project.

b. Land acquisition flood. A hypothetical or recorded flood event used to determine requirements for real estate acquisition.

c. Full pool. The maximum reservoir elevation for storing water for allocated project purposes.

d. Induced surcharge. Storage created in a reservoir above the top of flood control pool by regulating outflows during flood events.

e. Envelope curve. A curve which connects the high points of intersection of preproject and postproject water-surface profiles.

f. Guide taking line. A contour line used as a guide for land acquisition in the reservoir area. (Also referred to as the guide contour line or guide acquisition line.)

18-3. Real Estate Acquisition Policies for Reservoirs

a. Basic policies. Basic policies and procedures related to the acquisition of lands for reservoirs are presented in ER 405-1-12. Paragraph 2-12 of ER 405-1-12 states that, "Under the Joint Policy the Corps will take an adequate interest in lands, including areas required for public access, to accomplish all the authorized purposes of the project and thereby obtain maximum public benefits therefrom." Paragraph 2-12.*a*(2) further states that land to be acquired in fee shall include, "lands below a guide contour line...established with a reasonable freeboard allowance above the top pool elevation for storing water for flood control, navigation, power, irrigation, and other purposes, referred to in this paragraph as "full pool" elevation. In nonurban areas generally, this freeboard allowance will be established to include allowances for induced surcharge operations plus a reasonable additional freeboard to provide for adverse effects of saturation, wave action and bank erosion."

b. Considering factors. Factors such as estimated frequency of occurrence, probable accuracy of estimates, and relocation costs will be taken into consideration. Where freeboard does not provide a minimum of 300 ft horizontally from the conservation pool, defined as the top of all planned storage not devoted exclusively to flood-control storage, then the guide acquisition line will be increased to that extent. In the vicinity of urban communities or other areas of highly concentrated developments, the total freeboard allowance between the full pool elevation and the acquisition line may be greater than prescribed for nonurban areas generally. Also, there should be sufficient distance to assure that major hazards to life or unusually severe property damages would not result from floods up to the magnitude of the SPF. In such circumstances, however, consideration may be given to easements rather than fee acquisition for select sections if found to be in the public interest. However, when the project design provides a high level spillway, the crest of which for economy of construction is considerably higher than the storage elevation required to regulate the reservoir design flood, the upper level of fee acquisition will normally be at least equal to the top elevation of spillway gates or crest elevation of ungated spillway, and may exceed this elevation if necessary to conform with other criteria prescribed.

18-4. Hydrologic Evaluations

a. Development of land acquisition flood. To establish a reasonable surcharge allowance above the top pool elevation, a land acquisition flood, which includes the effects of any upstream reservoirs, should be selected and routed through the project to determine the impact on the establishment of the guide acquisition line.

b. Nonurban areas. In nonurban areas, the land acquisition flood should be selected from an evaluation of a range of floods with various frequencies of occurrence. The impact of induced surcharge operations on existing and future developments, hazards to life, land use, and relocations must be evaluated. The land acquisition flood will be chosen based on an evaluation of the risk and uncertainty associated with each of these frequency events. Basic considerations to be addressed during the land acquisition flood selection process should include the credibility of the analysis, identification and significance of risk, costs and benefits, and legal, social, and political ramifications.

c. Urban areas. In urban areas or other areas with highly concentrated areas of development, the SPF will be used for the land acquisition flood.

d. Project design sedimentation volume. Project capacity data should be adjusted for projected sediment volumes when routing the land acquisition flood. Project design sediment should be based on appropriate rates of sedimentation for the project area for the life of the project.

18-5. Water Surface Profile Computations

a. Backwater model development. A basic backwater model should be developed for the project area from the proposed flat pool area through the headwater area where impacts of the proposed reservoir are expected to be significant. The model should reflect appropriate cross-sectional data and include parameters based on historical flood discharges and high water marks. EM 1110-2-1416 presents the model requirements and calibration procedures.

b. Preproject profiles. A series of preproject water surface profiles should be developed utilizing preproject cross-section geometry, calibrated Manning's "n" values, and appropriate starting water surface elevations for the initial cross section. Flow rates used in the water surface profile computations should be selected from the peak and recession side of the land acquisition flood hydrograph.

c. Postproject profiles. A series of water surface profiles shall be developed utilizing the postproject cross sections which are adjusted to reflect project design sedimentation over the life of the project. Manning's roughness coefficients are based on adjusted preproject roughness coefficients to account for factors such as vegetation and land use changes which decrease hydraulic conveyance. Agricultural lands existing in the headwater areas prior to land purchases will likely revert to forested areas some years after the reservoir is filled. Preproject flow rates and coincident reservoir pool elevations from land acquisition flood routing should be used to compute postproject profiles.

d. Project design sedimentation distribution. Postproject cross-section geometry must be adjusted to reflect the impacts of sedimentation over the life of the project. Sedimentation problems associated with reservoir projects and methods of analysis to address sediment volumes and distributions are given in Chapter 5 of EM 1110-2-4000.

18-6. Development of an Envelope Curve

The development of an envelope curve is based on preproject and postproject water-surface profiles. A selected discharge from the land acquisition flood is used to compute a preproject and a postproject profile. A point of intersection is established where the profiles are within 1 ft of each other. The point of intersection is placed at the elevation of the higher of the two profiles. A series of points of intersection are derived from water-surface profile computations utilizing a range of selected discharges from the land acquisition flood. A curve is drawn through the series of points of intersection to establish the envelope curve.

18-7. Evaluations to Determine Guide Taking Lines (GTL)

a. Land acquisition flat pool. The land acquisition flat pool of a reservoir project is established by the maximum pool elevation designated for storing water for allocated project purposes to include induced surcharge storage and is not impacted by the backwater effects of main stream or tributary inflows. In flat pool areas, the elevations of the GTL are based on the flat pool elevation and a freeboard allowance to account for adverse effects of saturation, bank erosion, and wave action.

b. Headwater areas. In headwater areas, the GTL may be based on the envelope curve elevations and appropriate allowances to prevent damages associated with saturation, bank erosion, and wave action.

c. Flood-control projects. The selection of an appropriate land acquisition flood for flood-control projects located in rural areas should be based on an elevation of a range of frequency flood events. The land acquisition flood selection for flood-control projects in rural locations must include regulation by upstream reservoirs and reflect postproject conditions which minimize adverse impacts within the project area resulting from induced flood elevations and duration of flooding. In highly developed areas along the perimeter of flood-control projects, the SPF should be used for land acquisition. An envelope curve can be developed from the land acquisition flood routings and water-surface profile computations for preproject and postproject conditions. The land acquisition GTL may be established from the envelope curve and appropriate allowances for reservoir disturbances.

d. Nonflood-control projects. Nonflood control projects may be any combination of purposes such as water supply, hydropower, recreation, navigation and irrigation. The land acquisition flood selection process for nonflood-control projects located in rural areas is based on an evaluation of a range of frequency floods and is used to determine postproject flood elevations and duration of flooding in the project area. As with flood-control projects, regulation of flows by upstream reservoirs must be incorporated in the development process. The land acquisition flood used to evaluate real estate acquisitions in rural areas should reflect postproject conditions which minimize adverse impacts. The land acquisition flood for developed areas should be the SPF. The maximum pool elevation designed for storing water for allocated project purposes is used in the development of the land acquisition flood routing. An envelope curve based on preproject and postproject water-surface profiles utilizing project design sedimentation and distribution should be developed. The envelope curve and appropriate allowances for reservoir disturbances may be used to establish the land acquisition GTL.

18-8. Acquisitions of Lands for Reservoir Projects

Land acquisition policies of the Department of the Army governing acquisition of land for reservoir projects is published in ER 405-1-12, Change 6, dated 2 January 1979. Paragraph is as follows:

Joint Land Acquisition Policy for Reservoir Projects.
The joint policies of the Department of the Interior and Department of the Army, governing the acquisition of land for reservoir projects, are published in the Federal

Register, dated 22 February 1962, Volume 27, page 1734. On July 1966, the Joint Policy was again published in 31, F.R. 9108, as follows:

> JOINT POLICIES OF THE DEPARTMENTS OF THE INTERIOR AND OF THE ARMY RELATIVE TO RESERVOIR PROJECT LANDS

"A joint policy statement of the Department of the Interior and the Department of the Army was inadvertently issued as a notice in 27 F.R. 1734. Publication should have been made as a final rule replacing regulations then appearing in 43 CFR part 8. The policy as it appears in 27 F.R. 1734 has been the policy of the Department of the Interior and the Department of the Army since its publication as a Notice and is now codified as set forth below.

Section

8.0	*Acquisition of lands for reservoir projects*
8.1	*Lands for reservoir construction and operation*
8.2	*Additional lands for correlative purposes*
8.3	*Easements*
8.4	*Blocking out*
8.5	*Mineral rights*
8.6	*Building*

Authority: The provisions of this Part 8 issued under Sec. 7, 32 Stat., 389, Sec. 14, 53 Stat. 1197, 43 U.S.C. 421, 389.

8.0 Acquisition of Lands for Reservoir Projects.
Insofar as permitted by law, it is the policy of the Departments of the Interior and of the Army to acquire, as a part of reservoir project construction, adequate interest in lands necessary for the realization of optimum values for all purposes including additional land areas to assure full realization of optimum present and future outdoor recreational and fish and wildlife potentials of each reservoir.

8.1 Lands for Reservoir Construction and Operation.
The fee title will be acquired to the following:

a) Lands necessary for permanent structures.

b) Lands below the maximum flowage line of the reservoir including lands below a selected freeboard where necessary to safeguard against the effects of saturation, wave action, and bank erosion and to permit induced surcharge operation.

c) Lands needed to provide for public access to the maximum flowage line, as described in Paragraph 1b, or for operation and maintenance of the project.

8.2 Additional Lands for Correlative Purposes. The fee title will be acquired for the following:

a) Such lands as are needed to meet present and future requirements for fish and wildlife as are determined pur suant to the Fish and Wildlife Coordination Act.

b) Such lands as are needed to meet present and future public requirements for outdoor recreation, as may be authorized by Congress.

8.3 Easements. Easements in lieu of fee title may be taken only for lands that meet all of the following conditions:

a) Lands lying above the storage pool,

b) Lands in remote portions of the project area,

c) Lands determined to be of no substantial value for protection or enhancement of fish and wildlife resources, or for public outdoor recreation,

d) It is to the financial advantage of the Government to take easements in lieu of fee title.

8.4 Blocking Out. Blocking out will be accomplished in accordance with sound real estate practices, for example, on minor sectional subdivision lines: and normally, land will not be acquired to avoid severance damage if the owner will waive such damage.

8.5 Mineral Rights. Mineral, oil and gas rights will not be acquired except where the development thereof would interfere with project purposes, but mineral rights not acquired will be subordinated to the Government's right to regulate their development in a manner that will not interfere with the primary purposes of the project, including public access.

8.6 Buildings. Buildings for human occupancy as well as other structures which would interfere with the operation of the project for any project purpose will be prohibited on reservoir project lands."

18-9. Acquisition of Lands Downstream from Spillways for Hydrologic Safety Purposes

a. General. A real estate interest will be acquired downstream of dam and lake projects to assure adequate security for the general public in areas downstream from spillways. Real estate interests must be obtained for downstream areas where spillway discharges create or significantly increase a hazardous condition.

b. Evaluation criteria. Combinations of flood events and flood conditions which result in a hazardous condition or increase the hazard from the preproject to postproject flood conditions are determined for areas downstream from the spillway. These combinations of flood events and flood conditions are identified as critical conditions.

c. Flood events and conditions. Flood events up to the magnitude of the spillway design flood are evaluated for preproject and postproject conditions for areas downstream from the spillway. Flood conditions to be analyzed include flooded area, depth of flooding, duration, velocities, debris, and erosion.

d. Hazardous and nonhazardous conditions. The imposed critical conditions are analyzed to determine if these conditions are hazardous or nonhazardous. Nonhazardous areas are characterized by the following criteria:

(1) Flood depths do not exceed 2 ft in urban and rural areas.

(2) Flood depths are essentially nondamaging to urban property.

(3) Flood durations do not exceed 3 hr in urban areas and 24 hr in agricultural areas.

(4) Velocities do not exceed 4 fps.

(5) Debris and erosion potential are minimal.

(6) Imposed flood conditions would be infrequent. The exceedance frequency should be less than 1 percent.

Appendix A
References

A-1. Required Publications

Flood Control Act of 1944.

National Dam Safety Act, Public Law 92-367.

Water Resources Development Act of 1976.

Water Supply Act of 1958.

ER 405-1-12
Real Estate Handbook

ER 1110-2-240
Water Control Management

ER 1110-2-1451
Acquisition of Lands Downstream from Spillways for Hydrologic Safety Purposes

ER 1110-8-2(FR)
Inflow Design Floods for Dams and Reservoirs

EM 1110-2-1201
Reservoir Water Quality Analyses

EM 1110-2-1406
Runoff from Snowmelt

EM 1110-2-1411
Standard Project Flood Determinations

EM 1110-2-1412
Storm Surge Analysis and Design Water Level Determinations

EM 1110-2-1414
Water Levels and Wave Heights for Coastal Engineering Design

EM 1110-2-1415
Hydrologic Frequency Analysis

EM 1110-2-1416
River Hydraulics

EM 1110-2-1417
Flood Run-off Analysis

EM 1110-2-1419
Hydrologic Engineering Requirements for Flood Damage Reduction Studies

EM 1110-2-1602
Hydraulic Design of Reservoir Outlet Works

EM 1110-2-1603
Hydraulic Design of Spillways

EM 1110-2-1701
Hydropower

EM 1110-2-2904
Design of Breakwater and Jetties

EM 1110-2-3600
Management of Water Control Systems

EM 1110-2-4000
Sedimentation Investigations of Rivers and Reservoirs

ETL 1110-2-335
Development of Drought Contingency Plans

ETL 1110-2-336
Operation of Reservoir Systems

A-2. Computer Program/Document References

Fread 1989
Fread, D. 1989. BREACH: An Erosion Model for Earthen Dam Failures, User's Manual, National Weather Service, Hydrologic Research Laboratory, Silver Spring, MD.

Hydrologic Engineering Center (HEC) 1971
Hydrologic Engineering Center (HEC). 1971. HEC-4 "Monthly streamflow simulation," User's Manual, U.S. Army Corps of Engineers, Davis, CA.

HEC 1982c
HEC. 1982c. HEC-5, "Simulation of flood control and conservation systems," User's Manual, U.S. Army Corps of Engineers, Davis, CA.

HEC 1982d
HEC. 1982d. HYDUR, "Hydropower analysis using flow-duration procedures," User's Manual, U.S. Army Corps of Engineers, Davis, CA.

HEC 1984
HEC. 1984. HMR52, "Probable maximum storm (eastern United States)," User's Manual, U.S. Army Corps of Engineers, Davis, CA.

HEC 1986
HEC. 1986. HEC-5(Q), "Simulation of flood control and conservation systems," Appendix on Water Quality Analysis, U.S. Army Corps of Engineers, Davis, CA.

HEC 1987a
HEC. 1987a. STATS, "Statistical analysis of time series data," Input Description, U.S. Army Corps of Engineers, Davis, CA.

HEC 1990b
HEC. 1990b. "Flood damage analysis package on the microcomputer," Installation and User's Guide, Training Document 31, U.S. Army Corps of Engineers, Davis, CA.

HEC 1990c
HEC. 1990c. HEC-1, "Flood hydrograph package," User's Manual, U.S. Army Corps of Engineers, Davis, CA.

HEC 1990d
HEC. 1990d. HEC-2, "Water surface profiles," User's Manual, U.S. Army Corps of Engineers, Davis, CA.

HEC 1991a
HEC. 1991a. HEC-PRM, "Prescriptive reservoir model, program description," User's Manual, U.S. Army Corps of Engineers, Davis, CA.

HEC 1992c
HEC. 1992c. HEC-FFA, "Flood flow frequency analysis," User's Manual, U.S. Army Corps of Engineers, Davis, CA.

HEC 1993
HEC. 1993. HEC-6, Scour and deposition in rivers and reservoirs," User's Manual, U.S. Army Corps of Engineers, Davis, CA.

HEC 1995a
HEC. 1995a. HEC-DSS, "Users guide and utility program manual," U.S. Army Corps of Engineers, Davis, CA.

HEC 1995b
HEC. 199b. UNET, "One-dimensional unsteady flow through a full network of open channels," User's Manual, U.S. Army Corps of Engineers, Davis, CA.

Lane 1990
Lane. 1990. LAST, "Applied stochastic techniques," User's Manual, U.S. Bureau of Reclamation, Denver, CO.

USACE 1972
USACE. 1972. SUPER, "Regulation simulation and analysis of simulation for a multi-purpose reservoir system," Southwestern Division, Dallas, TX.

USACE 1991
USACE. 1991. SSARR, "Model streamflow synthesis and reservoir regulation," User's Manual, North Pacific Division, Portland, OR.

A-3. Related Publications

Alley and Burns 1983
Alley, W. M., and Burns, A. W. 1983. "Mixed station extension of monthly streamflow records," *Journal of Hydraulic Engineering*, ASCE, HY109(10), October 1272-1284.

Bowen 1987
Bowen, T. H. 1987. "Branch-bound enumeration for reservoir flood control plan selection," Research Document 35, HEC, Davis, CA.

Buchanan and Somers 1968
Buchanan, T. J., and Somers, W. P. 1968. "Stage measurements at gauging stations," U.S. Geological Survey, TWI 3-A7.

Buchanan and Somers 1969
Buchanan, T. J., and Somers, W. P. 1969. "Discharge measurements at gauging stations," U.S. Geological Survey, TWI 3-A8.

Carter and Davidian 1968
Carter, R. W., and Davidian, J. 1968. "General procedure for gauging streams," U.S. Geological Survey, TWI 3-A6.

DeVries and Hromadka 1993
DeVries, J. J., and Hromadka, T. V. 1993. "Computer models for surface water," *Handbook of hydrology* D. R. Maidment, ed., McGraw-Hill, New York.

Domenico and Schwartz 1990
Domenico, P. A., and Schwartz, F. W. 1990. *Physical and Chemical Hydrogeology*. John Wiley and Sons, New York.

Eichert and Davis 1976
Eichert, B. S., and Davis, D. W. 1976. "Sizing flood control reservoir systems by systems analysis," Technical Paper 44, Hydrologic Engineering Center, Davis, CA.

Farnsworth, Thompson, and Peck 1982
Farnsworth, R. K., Thompson, E. S., and Peck, E. L. 1982. "Evaporation atlas for the contiguous 48 United States," Technical Report NWS 33, National Oceanic and Atmospheric Administration (NOAA), Washington, DC.

Froelich 1987
Froelich, D. C. 1987. "Embankment-dam breach parameters," *Proceedings of the 1987 National Conference on Hydraulic Engineering at Williamsburg, VA.* American Society of Civil Engineers, New York.

Hoffman 1977
Hoffman, C. J. 1977. "Design of spillways and outlet works," *Handbook of Dam Engineering* A. R. Golzé, ed., Van Nostrand Reinhold, New York.

Hydrologic Engineering Center (HEC) 1977
Hydrologic Engineering Center (HEC). 1977. "Guidelines for calculating and routing a dam break flood," Research Document 5, U.S. Army Corps of Engineers, Davis, CA.

HEC 1980
HEC. 1980. "Flood emergency plan guidelines for Corps dams," Research Document 13, U.S. Army Corps of Engineers, Davis, CA.

HEC 1982a
HEC. 1982a. "Emergency planning for dams, bibliography and abstracts of selected publications," Research Document 17, U.S. Army Corps of Engineers, Davis, CA.

HEC 1983a
HEC. 1983a. "Example emergency plan for Blue Marsh Dam and Lake," Research Document 19, U.S. Army Corps of Engineers, Davis, CA.

HEC 1983b
HEC. 1983b. "Example plan for evacuation of Reading, Pennsylvania, in the event of emergencies at Blue Marsh Dam and Lake," Research Document 20, U.S. Army Corps of Engineers, Davis, CA.

HEC 1985a
HEC. 1985a. "Flood-damage-mitigation plan selection with branch-and-bound enumeration," Training Document 23, U.S. Army Corps of Engineers, Davis, CA.

HEC 1985b
HEC. 1985b. "Stochastic Analysis of Drought Phenomena," Training Document 25, U.S. Army Corps of Engineers, Davis, CA.

HEC 1990a
HEC. 1990a. "Modifying reservoir operations to improve capabilities for meeting water supply needs during droughts," Research Document 31, U.S. Army Corps of Engineers, Davis, CA.

HEC 1990e
HEC. 1990e. "A preliminary assessment of Corps of Engineers' reservoirs, their purposes and susceptibility to drought," Research Document 33, U.S. Army Corps of Engineers, Davis, CA.

HEC 1991b
HEC. 1991b. "Optimization of multiple-purpose reservoir system operations: A review of modeling and analysis approaches," Research Document 34, U.S. Army Corps of Engineers, Davis, CA.

HEC 1991c
HEC. 1991c. "Importance of surface-ground water interaction to Corps total water management: Regional and national examples," Research Document 32, U.S. Army Corps of Engineers, Davis, CA.

HEC 1991d
HEC. 1991d. "Missouri River System Analysis Model - Phase I," Project Report 15, U.S. Army Corps of Engineers, Davis, CA.

HEC 1991f
HEC. 1991f. "Columbia River System Analysis Model - Phase I," Project Report 16, U.S. Army Corps of Engineers, Davis, CA.

HEC 1992a
HEC. 1992a. "Missouri River System Analysis Model - Phase II," Project Report 17, U.S. Army Corps of Engineers, Davis, CA.

HEC 1992b
HEC. 1992b. "Developing Operation Plans from HEC Prescriptive Reservoir Model, Results for the Missouri System: Preliminary Results," Project Report 18, U.S. Army Corps of Engineers, Davis, CA.

HEC 1993
HEC. 1993. "Review of GIS Applications in Hydrologic Modeling," Technical Paper 144, U.S. Army Corps of Engineers, Davis, CA.

Linsley et al. 1992
Linsley, R. K., Franzini, J. B., Freyberg, D. L., and Tchobanoglous, G. 1992. *Water resources engineering*, 4th ed., McGraw-Hill, New York.

MacDonald and Langridge-Monopolic 1984
MacDonald, T. C., and Langridge-Monopolis, J. 1984. "Breaching characteristics of dam failures," *Journal of Hydraulic Engineering*, American Society of Civil Engineers, 110 (5), 567-586.

Matalas and Langbein 1962
Matalas, N. C., and Langbein, W. 1962. "Information Content of the Mean," in Journal of Geophysical Research, 67(9), 3441-3448.

Mays and Tung 1992
Mays, L. W., and Tung, Y-K. 1992. *Hydrosystems engineering & management*, McGraw-Hill, New York.

McGhee 1991
McGhee, T. J. 1991. *Water supply and sewerage*, 6th Edition, McGraw-Hill, New York.

Mosley and McKerchar 1993
Mosley, M. P., and McKerchar, A. I. 1993. "Streamflow," *Handbook of Hydrology*. D. R. Maidment ed., McGraw-Hill, New York.

National Weather Service 1960
National Weather Service. 1960. "Generalized estimates of PMP for the U.S. west of the 105th Meridian for areas less than 400 square miles and durations to 24 hr," Technical Paper 38, Silver Spring, MD.

National Weather Service 1977
National Weather Service. 1977. "Probable maximum precipitation estimates," United States east of 105th Meridian, Hydrometeorological Report No. 51, Silver Spring, MD.

National Weather Service 1981
National Weather Service. 1981. "Application of probable maximum precipitation estimates - United States east of the 105th meridian," Hydrometeorological Report No. 52, Silver Spring, MD.

Pruitt 1990
Pruitt, W. O. 1990. *Evaluation of Reservoir Evaporation Estimates*, submitted to USACE, Sacramento District, Sacramento, CA.

Salas 1992
Salas, J. D. 1992. "Analysis and modeling of hydrologic time series," *Handbook of Hydrology*, Maidment, ed.

Salas, Delleur, Yevjevich, and Lane 1980
Salas, J. D., Delleur, J. W., Yevjevich, V., and Lane, W. L. 1980. "Applied modeling of hydrologic time series," Water Resources Publications, Littleton, CO, 484.

Smoot and Novak 1969
Smoot, G. F., and Novak, C. E. 1969. "Measurement of discharge by moving-boat method," U.S. Geological Survey, TWI 3-A11.

Tasker 1983
Tasker, G. D. 1983. "Effective record length of T-year event," *Journal of Hydrology*, 64, 39-47.

Thomas and McAnally 1985
Thomas, W. A., and McAnally, W. H. 1985. "Open-Channel Flow and Seidmentation TABS2," Instruction Report HL-85-1, U.S. Army Engineer Waterways Experiment Station, Vicksburg, MS.

USACE 1962
USACE. 1962. "Waves in Inland Reservoirs: Summary Report on CWI Projects," CW-164 and CW-165. Washington, DC.

USACE 1979
USACE. 1979. National hydroelectric power study, Institute of Water Resources, Washington, DC.

USACE 1989
USACE. 1989. Digest of water resources policies and authorities, 1165-2-1, Washington, DC.

USACE 1990
USACE. 1990. "Stability of flood control channels," EC 1110-8-1(FR), Washington, DC.

USACE 1992
USACE. 1992. "Authorized and operating purposes of Corps of Engineers' reservoirs," Washington, DC.

U.S. Geological Survey 1992
United States Geological Survey. 1992. "Sediment deposition in U.S. Reservoirs - summary of data reported 1981-85," Interagency Advisory Committee on Water Data - Subcommittee on Sedimentation.

Waide 1986
Waide, J. B. 1986. "General guidelines for monitoring contaminants in reservoirs," Instruction Report E-86-1, U.S. Army Engineers Waterways Experiment Station, Vicksburg, MS.

Water Resources Council 1973
Water Resources Council. 1973. "Principles and standards for planning water and related land resources," *The Federal Register*, The National Archives of the United States, September 10, 1973.

Wurbs 1991
Wurbs, R. A. 1991. "Optimization of multiple-purpose reservoir system operations: A review of modeling and analysis approaches," Research Document 34, Hydrologic Engineering Center, Davis, CA.